MORE INNOVATIONS FOR DEVELOPMENT

Edited by Gillis Een and Sten Joste

Intermediate Technology Publications
in association with the Innovations for
Development Association 1991

Published by ITDG Publishing
The Schumacher Centre for Technology and Development
Bourton Hall, Bourton-on-Dunsmore, Rugby, Warwickshire CV23 9QZ, UK
www.itdgpublishing.org.uk

First published in 1991
Print on demand since 2004

ISBN 1 85339 102 6

A catalogue record for this book is available from the British Library

ITDG Publishing is the publishing arm of the Intermediate Technology Development Group. Our mission is to build the skills and capacity of people in developing countries through the dissemination of information in all forms, enabling them to improve the quality of their lives and that of future generations.

Printed in Great Britain by Lightning Source, Milton Keynes

TABLE OF CONTENTS

FOREWORD

The 1986 International Inventors Awards (IIA) resulted in a book named '100 INNOVATIONS FOR DEVELOPMENT'. It has been published in two editions - one in Sweden by IIA, and one in Britain by the Intermediate Technology Development Group (ITDG), which has been widely distributed by Oxfam. The book has been favorably reviewed, and we feel that it has served its purpose of disseminating information on Appropriate Technology also to those remote places where it is most needed.

The original IIA organization has been slightly modified and has changed its name into Innovation for Development Association or IDEA. We have received support from the Sven and Dagmar Salén Foundation and SIDA, which has made it possible for us to give five awards in 1990, and to plan for further awards. The 1990 IDEA Awards have been given to prominent innovators in five target areas i.e. Water, Energy, Forestry, Farming and Fishing. Of these, the latter two are new. Farming covers the whole food chain except aquaculture, which falls under Fishing.

We have received many nominations for the 1990 IDEA awards. The average quality has been very high this time, and we feel confident that the present book will turn out to be as useful as the previous one.

Technology Transfer

The main purpose of this new book is to disseminate knowledge of innovations which have proven their value in their present environments, and thus have the potential of becoming useful also in other parts of the world.

Each innovation has been described on one full page, with or without an illustration, or on half a page. The descriptions are far from complete, but hopefully sufficiently detailed to make it possible for a reader to decide whether he is interested enough to ask for more information.

A lot of effort from our side has gone into checking and supplementing addresses, telephone numbers etc. We feel that this is a very important part of this book. A potential user of the technology is thus able to contact the innovator and ask for more information, which hopefully will lead to a technology transfer of some kind.

We all agree that information about appropriate technology to Developing Countries is highly desirable. Much has been achieved, but the agencies involved have been criticized for directing a one way flow of not so appropriate technology from Developed Countries to Developing Countries.

We feel that what IDEA has achieved is a stimulation of technology transfer between Developing Countries. Most of the innovations that we describe in this book have been developed and tested in a real environment with a real need, and in close contact with the end users.

Criteria

In all fields of development it is important to start from a clearly identified and well defined need of some kind. The development has been successful when it has been able to satisfy that particular need.

It has, of course, been impossible for us to go very deep into all the projects that we have described in this book. We have, however, gone into considerable detail with the short-list that led to the naming of award winners.

Right from the start of the IDEA 1990 project we declared which criteria we were going to use for judging the nominations.They were:

- Sustainability: The long time use of natural resources without over-exploiting or exhausting them.

- Self-reliance: The independence from inputs of chemicals or energy, making it easier to survive within an unreliable infra-structure.

- Socio-economic acceptance: The invention is fully accepted by all those concerned, both from social and economic points of view.

We have favoured the systems approach. The result has been that many of the best nominations involve more than one target area. Good examples of this are the many interesting innovations in the field of agro-forestry.

Gillis Een *Sten Joste*

FARMING AND FOOD

Farming is the biggest of the target areas for the IDEA 1990 award. It covers not only the land-based primary production, but also the whole food chain up to, but not including, the cooking and preparation of the meals. Thus farming also includes processing and preservation of food. Farming received a larger number of nominations than any other target area. The quality of the nominations was high, and a majority of them met the criteria for appropriateness that we had defined from the outset, and among them a number of trends can be identified.

The prize in farming was given to Dr T.R. Preston for his work in animal husbandry. He has developed a system for more efficient feeding of ruminants, but also a system for intensive livestock production from locally available agricultural byproducts.

Many nominations pointed to ways of moving away from chemical farming, towards ecological farming. The rationale behind this is not only short-term economical gain, but also a development towards a Low Input Sustainable Agriculture (LISA) and subsequent self-reliance. One key issue is Biological Nitrogen Fixation (BNF). Many nominations suggested the use of annual or perennial plants of the pea family (*Leguminosae*), the goal being a LISA with a sustainable high output of proteins and other nitrogenous food products and components.

Another trend is the move away from monoculture, with its risk for pests, diseases and intolerable variations in market prices. Well integrated polycultures are much more stable and reliable. Many nominations could be labelled as agro-forestry, i.e. integrated systems of agriculture with supporting species of perennial trees or bushes, which supply support, protection against wind, shadow and BNF.

A third trend is the use of endemic genes from the local flora, which are bred into crop species, in order to develop new strains with better properties regarding resistance against pests and diseases, as well as tolerance against salt and drought.

Some nominations could be described as biological pest control, especially insect control, involving the import and breeding of such agents as viruses, bacteria and predatory wasps.

Gillis Een

INTENSIVE LIVESTOCK PRODUCTION FROM SUGAR CANE AND FORAGE TREES

PRIZE WINNER

Dr T. R. Preston
CIPAV
Apartado Aero 7482
Cali, COLOMBIA

A system for intensive livestock production from sugar cane and forage trees, targeted at resource-poor farmers in the tropics.

A farming system, developed in the Cauca Valley in Colombia, supports extremely high levels of livestock production, (of the order of 3,000kg of meat/ha/year), derived from environmentally protective perennial crops (sugar cane and nitrogen-fixing trees).

The sugarcane stalk, after removal of the tops, is fractionated into juice and bagasse, using a simple animal-powered 3-roll mill. The tree foliage is separated into leaves and wooden stems. The cane juice is a complete replacement for cereal grains and is the basis (75%) of a high quality diet for pigs. The cane tops are fed to African hair sheep. The tree leaves provide protein for both pigs and sheep. The bagasse and tree stems are used for fuel. The pigs and sheep are confined and the excreta recycled through plastic bag biogas digesters, ponds with fish and aquatic plants, and through earth worm cultivation.

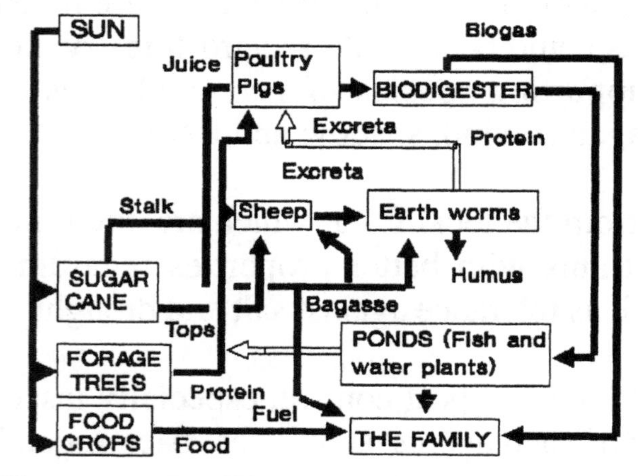

These elements provide additional household fuel, protein for the livestock and fertilizers for the crops.

More than one thousand farmers in Colombia are using some or all of the elements that make up this system.

Contact: Dr Thomas Reginald Preston, Convenio Inter-institucional para la Produccion Agropecuaria en el Valle del Rio Cauca (CIPAV), Calle 8a, No 3-14, Piso 9, Apartado Aero 7482, Cali, COLOMBIA, telex 055724, telephone +57-23-823271 ext. 328-444, telefax +57-23-824627.

RUMINANT FEEDING SYSTEM

PRIZE WINNER

Dr T. R. Preston
CIPAV
Apartado Aero 7482
Cali, COLOMBIA

A system for feeding of ruminants with emphasis on a proper balance of essential nutrients in the products of digestion.

This innovation is based on the discovery of the 'bypass protein' principle in feeding ruminants. Certain forms of protein, such as heat-treated fish meal, escape the rumen fermentation and contribute amino acids directly in the small intestine. The resulting feeding system uses molasses, urea and fish meal as basic components and was first applied in sugarcane growing countries, such as Cuba, Mauritius and Mexico.

Later on, local alternatives to fish meal were developed. One of these is cereal milling byproducts, such as rice polishings. Another example is leaves from nitrogen-fixing trees, such as *Leucaena leucocephala.*

A second discovery led to the replacement of some sugar in the feed with starch, which also has 'bypass' properies.

These discoveries led to successful commercial applications in tropical countries, where the feed resources, derived from crop residues and agro-industrial by-products, otherwise tend to be out of balance.

Effect of bypass protein (fish meal) on performance of steers fed a basal diet of molasses-urea

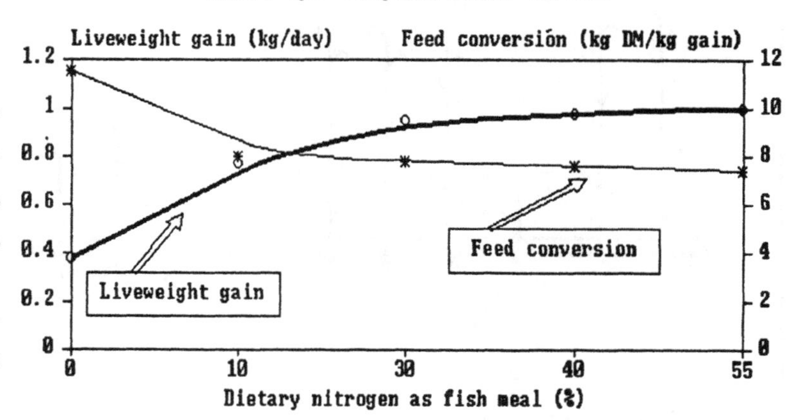

Contact: Dr Thomas Reginald Preston, Convenio Inter-institucional para la Produccion Agropecuaria en el Valle del Rio Cauca (CIPAV), Calle 8a, No 3-14, Piso 9, Apartado Aero 7482, Cali, COLOMBIA, telex 055724, telephone +57-23-823271 ext. 328-444, telefax +57-23-824627.

ALLEY CROPPING

HONORARY AWARD WINNER

Dr Biauw Tjwan Kang
IITA
P.M.B. 5320, Ibadan
NIGERIA

A system for sustainable small scale farming in humid and subhumid tropical areas.

Alley cropping is a sustainable form of agro-forestry, which offers an opportunity for intensive cultivation, and for saving humid and subhumid tropical forests from total destruction. The system involves integration of multipurpose trees and shrubs, which are planted in rows. Food crops are grown in the alleys between the hedgerows.

The trees and shrubs have the following functions in the system: To stabilize the soil against water runoff and soil erosion; To fix atmospheric nitrogen; To provide green manure or mulch for the food crops; To recycle mineral nutrients from deep soil layers; To provide protein-rich feed for small household animals; To give shade and wind protection; To give physical support to climbing crops; To produce household fuel.

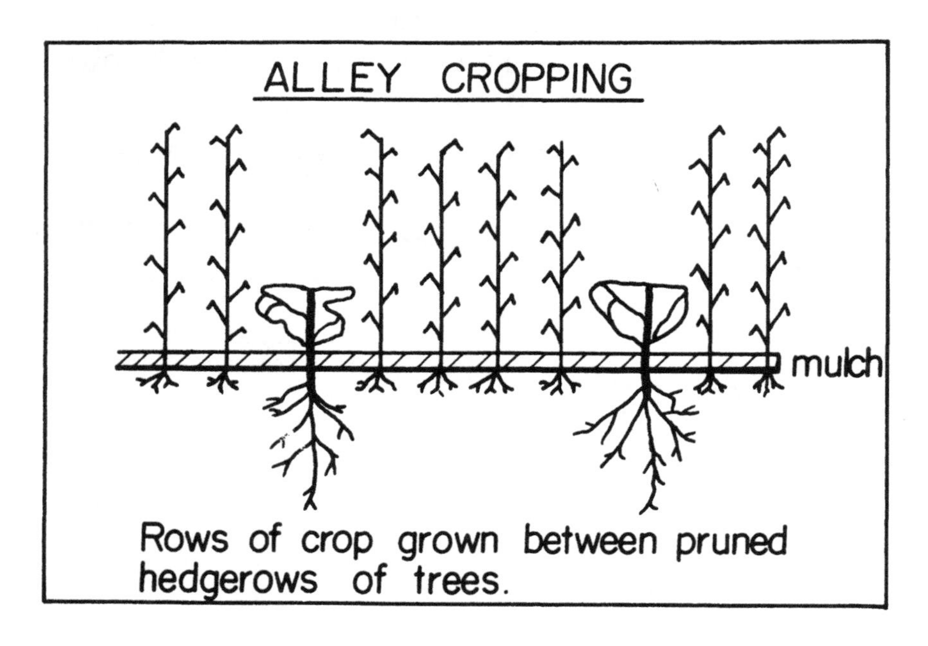

Rows of crop grown between pruned hedgerows of trees.

The system offers many advantages, such as: Longer cropping periods; Rapid soil-fertility regeneration; Soil and water conservation; Reduced external inputs such as chemical fertilizers; Suitability for large as well as small farms.

Contact: Dr Biauw Tjwan Kang, International Institute of Tropical Agriculture (IITA), P.M.B. 5320, Ibadan, NIGERIA, telex 31417 TROPIB NG.

POTATO PROCESSING AT VILLAGE LEVEL

Mr Robert Nave
SOTEC
Post Box 75
Bareilly 243001
U.P., INDIA

Several processes for storage and upgrading of locally grown crops, such as potato, soya, vegetables and fruit. At present dried potato products and flour are produced.

A demonstration centre - called the Research, Training and Village Development Centre (RTVDC) - has been established by SOTEC, for the purpose of developing appropriate processes for storage and upgrading of locally grown crops. The potato slicing device is one example of what has been produced.

The potato slicer consists of a cutting head with one adjustable blade, an impeller, an opening through which potatoes are fed and a tank with water for lubrication. The machine is driven by a cycle frame with pedals, (or by an electric motor). The throughput is 200-300 (300-600) kgs/hr of potatoes.

The slicer has been developed in co-operation with the International Potato Center (CIP), Region 4, IARI Campus, New Delhi 110012, INDIA, and with Compatible Technology Inc. (CTI), 5835 Lyndale Avenue South, Minneapolis, MN 55419, USA.

Contact: Robert W. Nave, Project Director, Society for Development of Appropriate Technology, (SOTEC), 182 Civil Lines (Jail Road), Post Box 75, Bareilly 243 001, Uttar Pradesh, INDIA, telephone +91-581-79049, telefax +91-581-72758, telex 577-227 GREL IN

SUNFLOWER OIL EXTRACTOR

Mr Carl Bielenberg
ATI
Washington, DC, USA

A manually operated press for the extraction of vegetable oil from sunflower seeds.

This sunflower seed press is manually operated, has continuous throughput, and is able to develop the high pressure needed for extracting oil from whole seeds. It is of the ram-type and the required pressure is produced by pulling a long lever arm, thus moving a piston through a compressing cage. The extraction rate is between 60 and 70 % of the available oil.

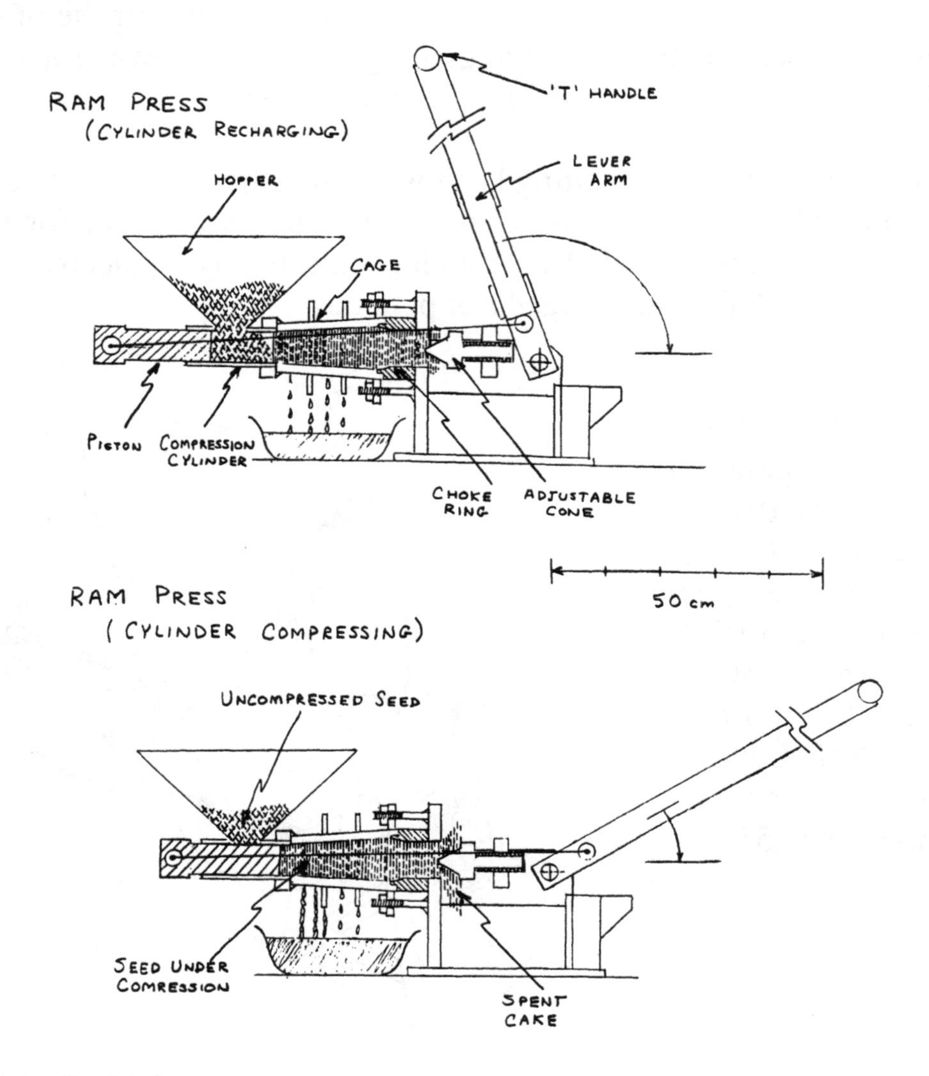

Contact: Mr Carl Bielenberg, c/o Appropriate Technology International (ATI), 1331 H Street NW, Washington, D.C. 20005, USA, telephone +1-202-879-2900, telefax +1-202-628-4622, telex 64661 ATI.

ANIMAL-DRAWN BROAD-BED-MAKER

Dr T. Mamo
Alemaya U.of Agric.
Debre Zeit, ETHIOPIA

A farm implement for the production of broad-beds-and-furrows (BBF) in African vertisoils.

The broad-bed-maker (BBM) has been developed primarily in order to make it possible to utilize better the Ethiopian vertisoil. This soil is very difficult to cultivate, and mismanagement can easily result in low yields as well as severe sheet and gully erosion.

The BBM is an animal-drawn farm implement, consisting of two joined ploughs and a chain. The two ploughs shift the soil towards the centre, and the chain works as a harrow. The BBM produces raised broad beds 1.2m wide and 0.2m high, commonly called broad-beds-and-furrows (BBF).

This innovation makes it possible to plant a first crop with the first rains, and also to take a second crop. The result is a 50% increase in grain and straw production, the latter being used as animal feed.

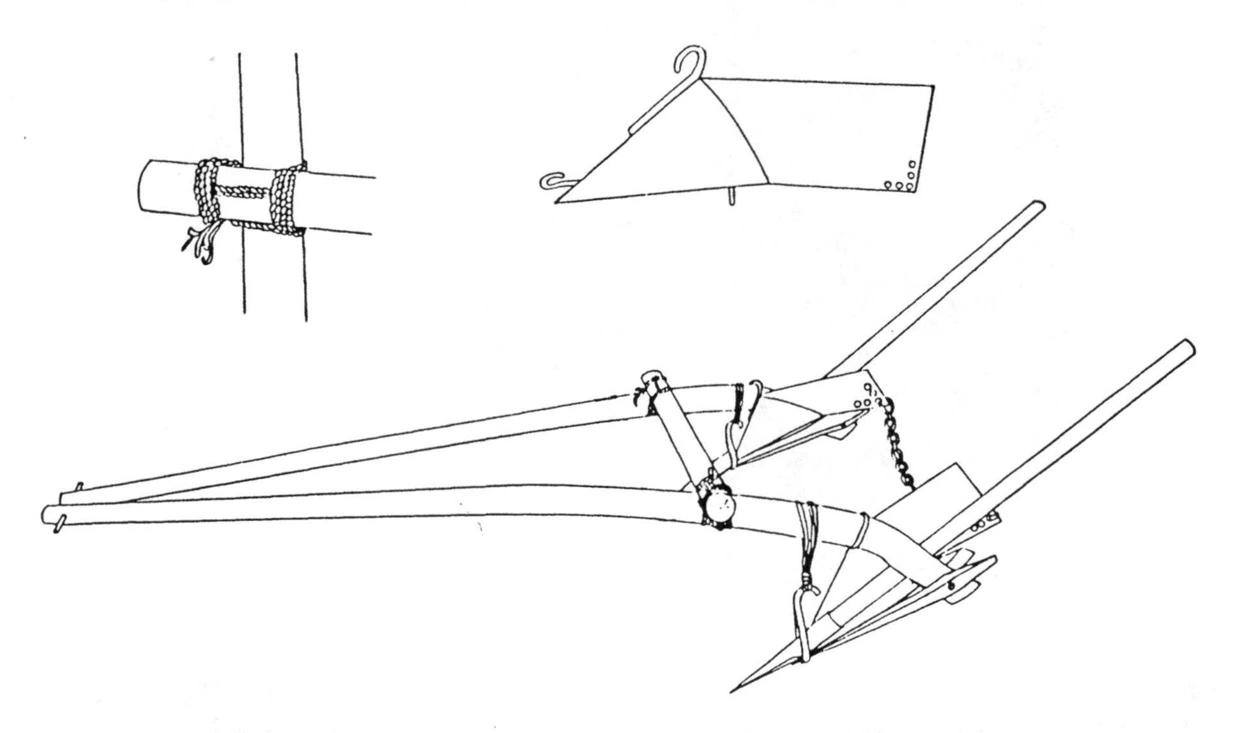

Contact: Dr Tekalign Mamo, Alemaya University of Agriculture, Debre Zeit Research Centre, Debre Zeit, ETHIOPIA.

MUSHROOM CULTIVATION

Dr Shu-Ting Chang
The Chinese University
HONG KONG

Cultivation of tropical mushrooms on agricultural or industrial solid waste materials.

A complete process for the indoor cultivation of tropical mushrooms on agricultural or industrial solid waste materials, such as straw or cotton waste. The substrate is first composted, then pasteurized with steam and finally inoculated. The waste after the last mushroom harvest is a good soil conditioner for growing vegetables.

The species utilized for this purpose are selected from the genuses *Pleurotus* and *Volvariella*, i.e. species which have a long tradition of being cultivated for food in the eastern parts of Asia. The technique is simple and requires small land areas. Thus it is suitable for rural regions with a dense population. Genetic manipulation has contributed to the development of fast growing varieties of tropical mushrooms. As little as nine days between spawning and harvesting has been achieved.

Pleurotus sajor-caju (Fr.) Sing. grown on cotton waste mixed with paddy straw compost

Contact: Professor Shu-Ting Chang, Department of Biology, The Chinese University of Hong Kong, Shatin, NT, Hong Kong, telephone +852-6952286, telefax +852-6035646, telex 50301 CUHK HX.

RECLAMATION OF SALINE AND ALKALINE SOILS

Dr I.P. Abrol
ICAR
New Delhi, INDIA

Crop production and silvi-pastoral systems, developed specifically for better utilization of saline and alkaline soils.

A large programme for the development and rehabilitation of salt affected soils. Among the methods now available, the following are the most important:

- Diagnostic criteria for salt affected soils.
- Salt tolerance has been established for several field crops.
- Alkali tolerance has been established for a large number of species of grasses, bushes and trees.
- A whole range of management practices have been defined in order to opti-mize production on affected soils.
- The 'Auger Hole Technique' for planting trees in alkaline soils.

These methods not only result in better soil productivity, but also in significant environmental improvement through reduced run-off, improved soil permeability etc. Furthermore the system provides wood and household fuel for the farmers.

Contact: Dr I.P. Abrol, Deputy Director General (Soils, Agronomy and Agroforestry), Indian Council of Agricultural Research (ICAR), Krishi Bhavan, Dr Rajendra Prasad Road, New Delhi 110 001, INDIA, telephone office +91-11-383762, telephone home +91-11-675381, telex 031-62249 ICAR IN.

EVAPORATIVELY COOLED POTATO STORE

Mr V. B. Sikka
78 Adarsh Nagar
Jalandhar 144008 PB
INDIA

A cooled store for keeping potatoes between the growing seasons. Water is evaporated by natural draught through moist beds of wood 'wool'.

A store for food potatoes in order to make possible marketing the year round, as well as for next season's seed potatoes. The cooling effect is from an outside air temperature of 40 degrees centigrade to an inside temperature of 20 to 25 degrees. This type of potato store is used by thousands of Indian potato growers.

The potato store-room consists of a closed and insulated room, that can hold about 250 bags (100kg each) of potatoes. Water is moved manually to a tank at a higher level, from where it trickles through beds of wood 'wool' suspended on chicken mesh. Draft is created by a black, and thus heat-catching, chimney and roof. Furthermore the air intake is directed towards the prevailing wind.

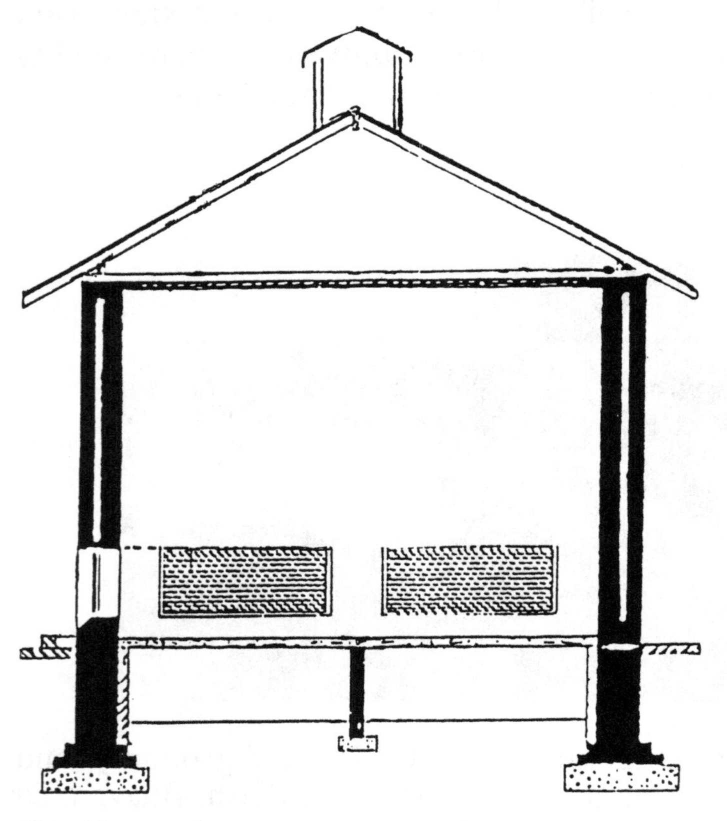

The inventor claims the following advantages:

- It is easy and cheap to construct.
- It can be built with locally available material.
- It does not need any electrical power.
- No trained mechanics or otherwise skilled people are required for the operation and maintenance.
- There are no running costs except labour.

Contact: Mr V. B. Sikka, Mehr Chand Polytechnic, Jalandhar City, Jalandhar 144 008 PB, Punjab, INDIA, telephone +91-181-72848, telefax + 91-181-73792.

BIOLOGICAL PEST CONTROL

Dr Flavio Moscardi
EMBRAPA/CNPSo
Londrina, PR, BRAZIL

The use of a nuclear polyhedrosis virus to control the velvetbean caterpillar on soya beans.

A method of biological pest control directed against the caterpillar of the insect species *Anticarsia gemmatalis*, which can do serious damage on soya beans. The active agent is a nuclear polyhedrosis virus with the acronym AgNPV. The method is specific and replaces broad spectrum chemical pesticides, which have the disadvantage of killing also the parasites and predators of the pest.

The virus is produced under laboratory or field conditions by CNPSo, as well as by some farmer co-operatives, and distributed to the soybean growers. After application, diseased and dead larvae are collected in the field, mixed with water,liquefied and filtered. The liquid is sprayed in the field at an appropriate stage of the development of the caterpillar. More than 90% of the virus produced this way and thereafter, stays in the soil, where a significant percentage of virus remains active until the following crop season.

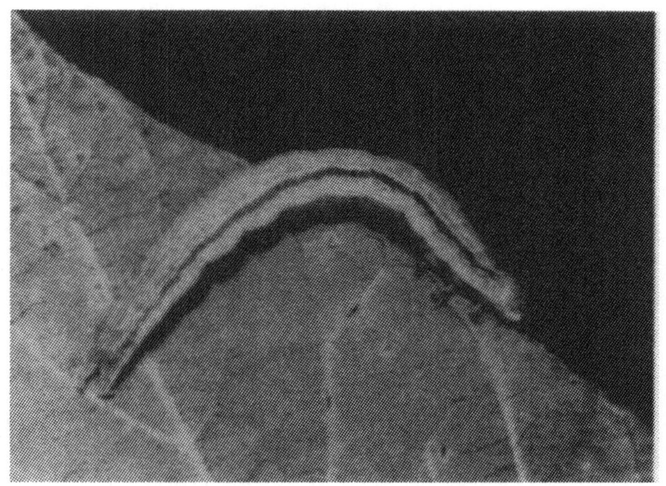

More sophisticated methods for propagation of the virus, formulation, storage between seasons and application have been developed, but these are applied mainly for start-up operations in new areas.

There are no patents, and the technology is free to use. At CNPSo there is a continuous selection of more virulent new wild strains of the same type of virus. Similar methods have been developed against *Erinnyis ello* (on cassava), *Spodoptera frugiperda* (on maize) and *Eacles imperialis magnifica* (on cashew).

Contact: Dr Flavio Moscardi, Empresa Brasileira de Pesquisa Agropecuária (EMBRAPA), Centro Nacional de Pesquisa de Soja (CNPSo), P.O.Box 1061, 86001 Londrina, Parana, BRAZIL, telephone +55-432-204166, telefax +55-432-204186, telex 432-208.

SHEA BUTTER EXTRACTION

Mr S. K. Adjorlolo
SIS Engineering Ltd.
Kumasi, GHANA

A plain and uncomplicated kneader to be used in the process of extraction of shea butter from shea nuts.

Mr Adjorlolo has developed a kneader to be used in a traditional method for the extraction of shea butter.

The shea nut tree grows wild in Central and West Africa. The fat, i.e. the shea butter, is traditionally extracted by means of slow and laborious manual methods.

The new kneader reduces the hand labour input to one tenth of that required with the traditional method. The complete process becomes faster and each person can produce more. Yield and quality of extracted fat remains the same as with the best of the old wet methods, but much better than with the dry methods.

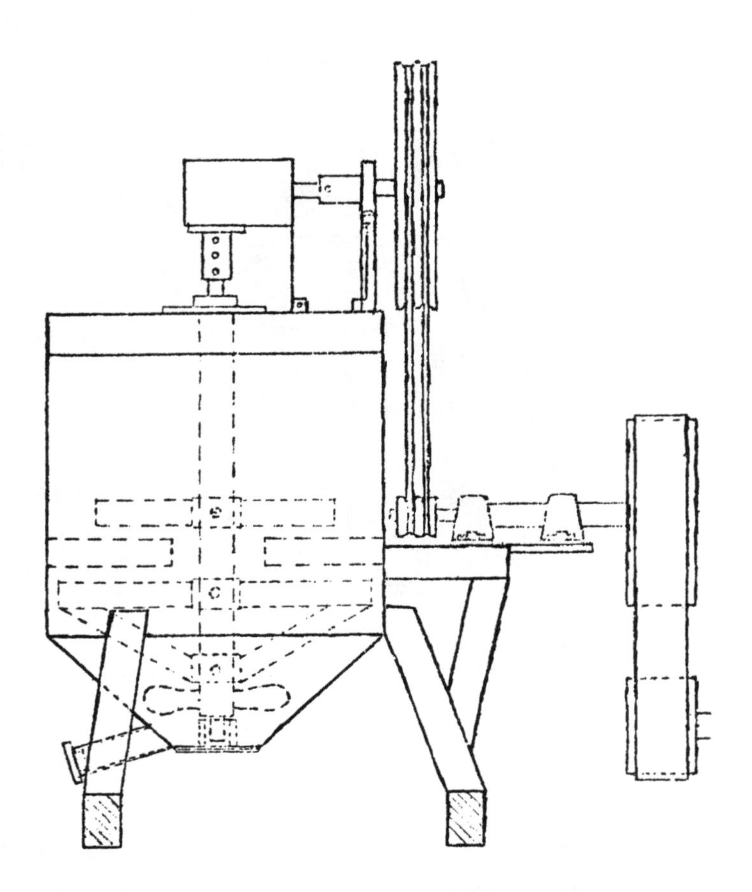

The shea butter extraction is by tradition carried out by women in the villages, who have now completely accepted the new kneader.

The kneader consists of a cylindrical tank with conical bottom. The shea kernel paste is poured into the tank and stirred continuously, while warm water is added.

Contact: Mr Solomon Kafui Adjorlolo, SIS Engineering Ltd., University P.O. Box 45, Carpenter Line, Oforikrom, Kumasi, GHANA.

GARRI-FRYING MACHINE

Edmund O. Kaine
PRODA
Enugu, Anambra State
NIGERIA

A frying and drying machine for garri. The last step in a continuous process for making food from cassava.

Garri (or gari) is made of cassava and is a staple food in many parts of West Africa. The process for making Garri is rather long and complicated, but every step is essential in order to produce an entirely safe food product. The two most important steps are a solid state fermentation and a frying / drying operation.

Mr Kaine has designed a machine for the last step in the garri process, i.e. the frying and drying of the final product. The cylindrical pan is stirred with spring-loaded blades attached to a horizontal shaft, driven by a small electrical motor. The pan is heated with diesel fuel. A sensor keeps the temperature between 135 and 145 degrees centigrade. The machine is continuous and the inlet end acts as a frying pan, while the outlet end is provided with vent holes and acts as a drier.

The garri produced in this machine has the the same high quality as any product made by hand, and has the wellknown shape of spherical grains. It can handle very large quantities if required, e.g. when the harvesting of cassava is at its height. Fresh cassava can not be stored, but processed Garri can. The machine is hygienic, easy to clean and smokeless.

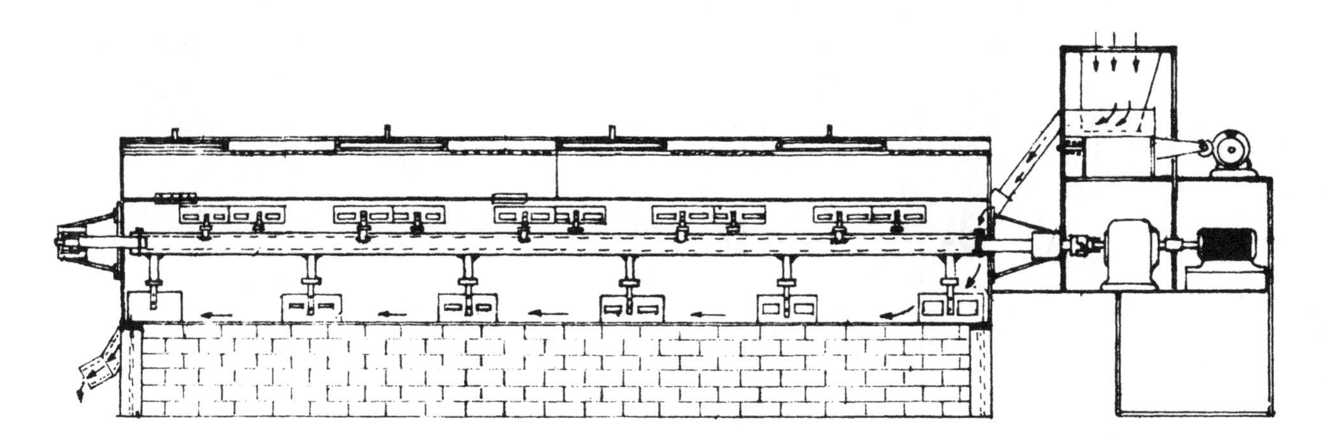

Contact: Mr Edmund O. Kaine, Projects Development Institute (PRODA), Federal Ministry of Science and Technology, P.O.Box 609, Enugu, Anambra State, NIGERIA, telephone +234-642-331593.

INTERGENERIC WHEAT HYBRIDS

Prof. Yushen Dong
CAAS
Beijing, CHINA

Development of disease-resistant intergeneric hybrids between wheat and native grass species.

Professor Dong has travelled extensively in China in order to collect germplasm of native varieties of a subspecies of wheat (*Triticum aestivum ssp. yunnanense*), as well as native species of grasses belonging to many different genera (*Agropyron, Aeguilops, Elymus, Elytrigia, Eremopyrum, Hordeum, Leymus, Psathyrostachys, Roegneria, Secale*).

大赖草 Leymus racemosus (Lam.) Tzvel.

Many amphiploids of *Triticum* and *Aegilops* have been developed. The native Chinese species *Aegilops desertorum, Aegilops michnoi, Leymus multicaulis, Leymus racemosus* and *Psathyrostachys juncea* have been successfully crossed with wheat. Several intergeneric hybrid lines, which come from these parents, are resistant to wheat powdery mildew and to barley yellow dwarf virus (BYDV).

Contact: Professor Yushen Dong, Institute of Crop Germplasm Resources, Chinese Academy of Agricultural Sciences (CAAS), Bai-Shi-Qiao Road No. 30, Beijing 100081, CHINA, telephone +86-1-831-4433-2726, telefax +86-1-831-6545, telex 222720 CAAS CN.

EARTHWORM TECHNOLOGY

Professor M.R. Bhiday
Erandwane
Pune 411004
INDIA

A complete system for the use of earthworms for the improvement of soil quality and crop yields in small-scale farming.

Professor Bhiday has developed a complete system for the employment of earthworms in small-scale agriculture. He has called the system 'Vermi-culture' or 'Vermi-compost'. The system consists of the following main steps:

1. Agricultural, animal and human waste material is collected and digested in anaerobic bioreactors. Also cellulosic material such as straw, paper and cardboard can be used. The resulting biogas is collected and burned locally as fuel in cooking stoves.

2. The sludge from the bioreactors is drained and used as substrate for the cultivation of selected varieties of earthworms. The resulting compost, together with its earthworms, is used as fertilizer and soil conditioner.

3. Earthworms have very good protein qualities and any surplus is fed to monogastric animals, or fish.

Vermi-compost results in better soil porosity, better water-holding capacity and thus less need for irrigation. Nutrients are transported vertically in the soil. The technology has proved its value in small-scale cultivation of sugarcane, fruits and vegetables, in re-forestation projects etc.

Contact: Professor M.R. Bhiday, 100 B Kalpana Apartments, Erandwane, Pune 411004, INDIA, telephone +91-212-336216.

CULTIVATION OF CACTUS FIGS

Dr Y. Gutterman
Inst. Desert Research
Ben-Gurion Univ., ISRAEL

A method of increasing the fruit yield of *Opuntia ficus-indica* under desert conditions, using low-quality irrigation water.

The fig-cactus or prickly pear (*Opuntia ficus-indica),* is a native of the American continent, but is now common in suitable places all over the world. The large yellow-orange fig-like fruit is edible and tasty. It has also been planted for pasture. The fig-cactus grows well in poor desert soil and can be irrigated with low-quality water. It can survive several months of complete drought.

The invention, which is adapted to the arid zones of the world, consists of taking cuttings in the form of terminal branches (leaves). The cuttings are planted in January, flowers profusely in May, to produce a good crop in July. A less spiny variety has been used.

The spent cuttings can be used as animal feed. If the crop is irrigated, the cuttings produce new stem parts, which can be used as a vegetable in e.g. sweet-sour Mexican dishes.

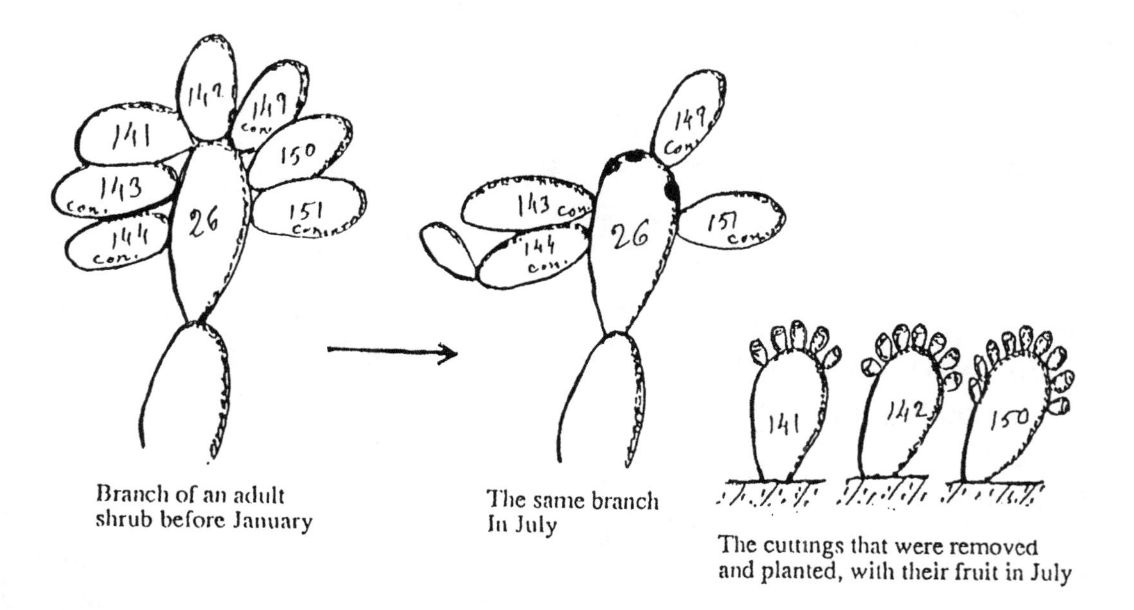

Branch of an adult shrub before January

The same branch In July

The cuttings that were removed and planted, with their fruit in July

Contact: Dr Yitzchak Gutterman, The Jacob Blaustein Institute for Desert Research, Ben-Gurion University of the Negev, Sede Boker Campus, ISRAEL 84990, telephone +972-57-565074, telex 5280 DIRBG IL, telefax +972-57-555058.

JAB-PLANTER

Mr Alfredo Zuniga
UDA
Comayagua, HONDURAS

A simple jab-planter for maize or beans, which can be adapted to local needs and customs, and which is 27% faster to use.

Mr Zuniga, in co-operation with the AFRC Institute of Engineering Research in UK, has developed a simple and efficient jab-planter for maize and beans.

The upper half of the traditional planting stick is replaced by a length of PVC tube, which acts as a hopper to store seeds. A metering slide, operated by a simple trigger, passes through the bottom of the hopper.

A seed, removed from the hopper, is caught by a funnel and delivered to the chisel point by a second piece of tube strapped to a wooden shaft. To the lower end of this, farmers like to fit their own steel points, which have evolved into a wide range of shapes to suit different soil types. The planter weighs 1.9 kilogrammes empty and can carry 0.26 kilogrammes of seed. This design is 27% faster to use than the traditional planter.

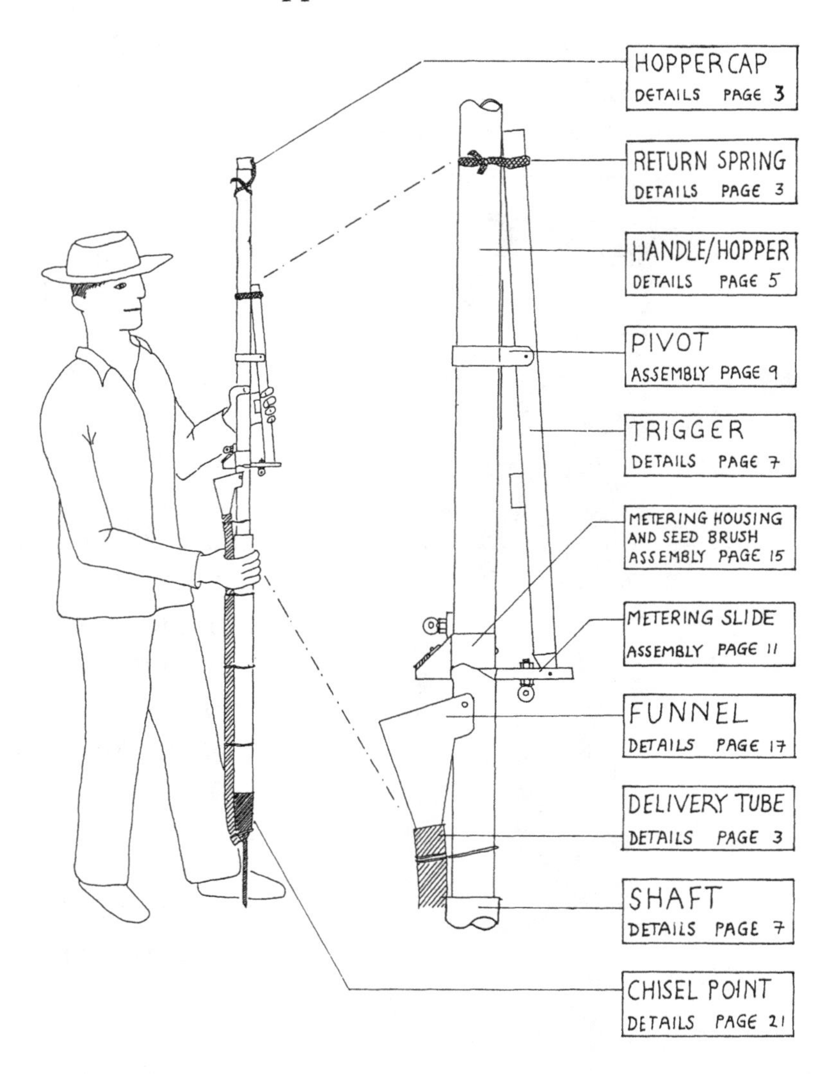

Contact: Mr Alfredo Zuniga, Unit of Development and Adaption (UDA), Ministry of Natural Resources, AP 133, Comayagua, HONDURAS.

MACHINE FOR CUPS & PLATES FROM LEAVES

Mr P. Veerraju
Mr J.K. Crown
CFTRI Mysore, INDIA

A simple machine for making cups and and plates for food from e.g. dried banana leaves.

In India cups and plates made of leaves are traditionally used for serving food and for packaging items such as butter and sweets. These are made from the broad leaves of banana, *Butea, Bauhinia* etc., and also from the leaf sheath of the areca palm.

The inventors have developed a simple machine for making such cups and plates from dried leaves. These are washed and moistened in order to make them pliable in preparation for the shaping process.

The machine is a simple, pedal-operated device. The upper part has a stationary male die, provided with electrical heating. It also holds a peripheral steel blade cutter. The lower part consists of a matching female die, provided with an aluminium cutting pad, and is moved by the pedal. The electricity supply is ordinary single phase 220V household energy, and the temperature is held at about 150 degrees centigrade by an energy regulator. The leaf is held in place, the lower die is pushed up with the pedal, and the object is formed. The heat dries the object and stabilizes its shape. Further pressing with the pedal results in the trimming off of excess material. The heating in the die also has the effect of sterilizing or sanitizing the object, thus destroying mould spores etc.

Contact: Mr P. Veerraju, Centre for Food Packaging, Central Food Technology Research Institute (CFTRI), Mysore 570 013, INDIA, telephone +91-821-37052, telex 846-241 FTRI IN.

DEEP RIPPING

Dr G.I. Nilsson
Sanitas Ltd.
Gaborone
BOTSWANA

25% of the farmland is ploughed in parallel strips as deep as possible. The effect is a much improved water economy.

Deep ripping of permanent strips is an agricultural technology that Dr Nilsson has developed for an area with scarce and irregular rainfall. It can be described as follows:

- About 25% of the land is ploughed or ripped in parallel strips, on sloping land along the contour.
- The ploughing or ripping goes as deep as is possible with the plough and traction available.
- The crops are planted in the ploughed strips, while the hard interstrip area is used for farm implements and transport.

The inventor claims the following advantages with the system:

- The deep rips collect more water when it rains and thus also prevent erosion.
- The plant roots grow deep into the soil and thus the plants become more resistant to drought conditions.
- Farm implements, with tractors and animals, move on the hard interstrip areas and thus do not compact the farmed strips.

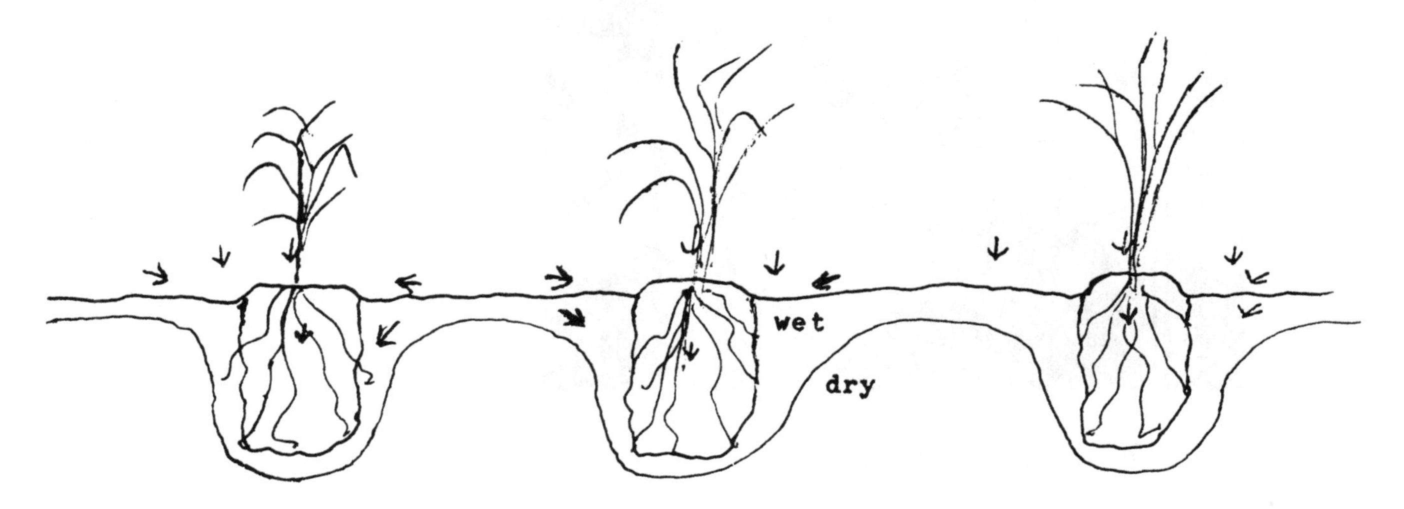

Contact: Dr G.I. Nilsson, Sanitas (Botswana) (Pty.) Ltd., Plant Clinic & Nurseries, P.O.Box 606, Gaborone, BOTSWANA, telephone +267-352538.

BEANS RESISTANT TO STORAGE PESTS

Dr César Cardona
CIAT
Cali
COLOMBIA

New varieties of dry beans have been developed, which are resistant in storage to the Mexican bean veevil, a serious tropical pest.

Dr Cardona has developed new varieties of dry beans with high levels of resistance to the Mexican bean weevil (*Zabrotes subfasciatus*). This insect is an important and cosmopolitan pest of beans in the tropical lowlands of Africa, Asia and Latin America.

This innovation holds great potential in those parts of the world where this insect causes severe damage to stored beans. Small farmers fear losses from the weevil and are forced to sell their bean crop promptly, even at times of surplus and low prices.

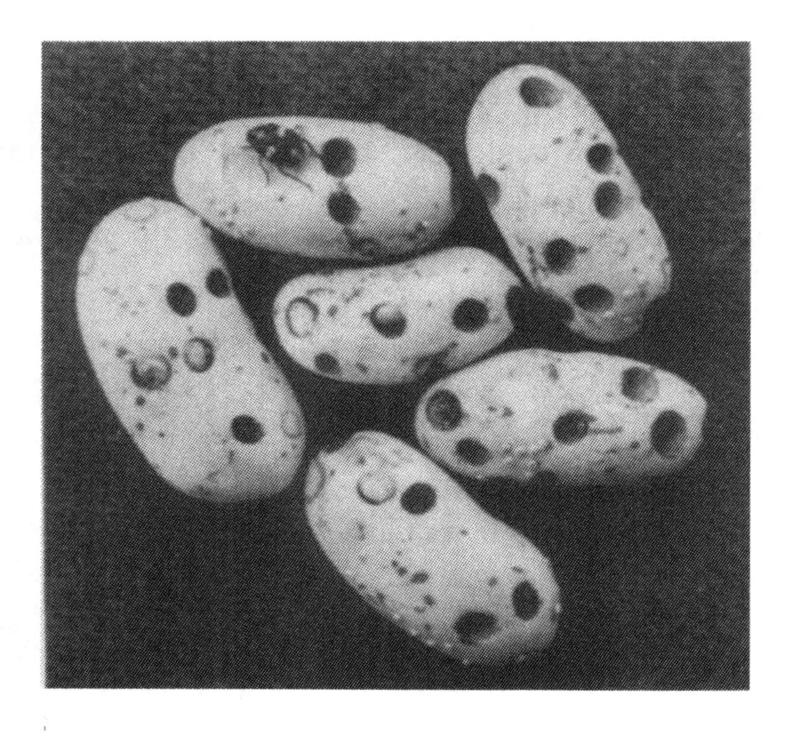

Weevil and damaged beans

Search among wild Mexican beans resulted in the discovery of varieties with a high natural resistance against the weevil. Research disclosed that this is due to a storage protein called *arcelin*. Thus it became possible to breed, in a relatively short time, new pest-resistant varieties of beans with good nutritional and agronomic characteristics

Contact: Dr César Cardona, Centro International de Agricultura Tropical (CIAT), Apartado aéreo 6713, Cali, COLOMBIA, telephone +57-23-675050, telefax +57-23-647243, telex 05769 CIAT CO.

BIOLOGICAL CONTROL of the CASSAVA MEALYBUG

Dr Hans R. Herren
IITA
Cotonou
BENIN

A parasitoid wasp was imported from South America, the native continent of the mealybug. The wasp is now firmly established in Africa.

The mealybug *Phenacoccus manihoti* is a serious pest on cassava (*Manihot esculenta*) in many parts of Africa, often reducing the crop yield by as much as 85%. The use of chemical pesticides is inappropriate, expensive and dangerous for the small farmer, who may risk poisoning himself, his family and his animals, as well as spoiling the environment.

Biological control is very appropriate in this case. It means the release of natural enemies to the pest, which have been fetched from their area of origin. The first success was the release and establishment of the parasitoid wasp *Epidinocarsis lopezi*, which was imported from South America, cassava's and the mealybug´s continent of origin. Both the adult wasp and its larvae feed on and kill the mealybug.

The wasp *Epidinocarsis lopezi* has now been released in over 150 sites in the cassava belt. The area controlled by the wasp covers approximately 2.7 million square kilometres.

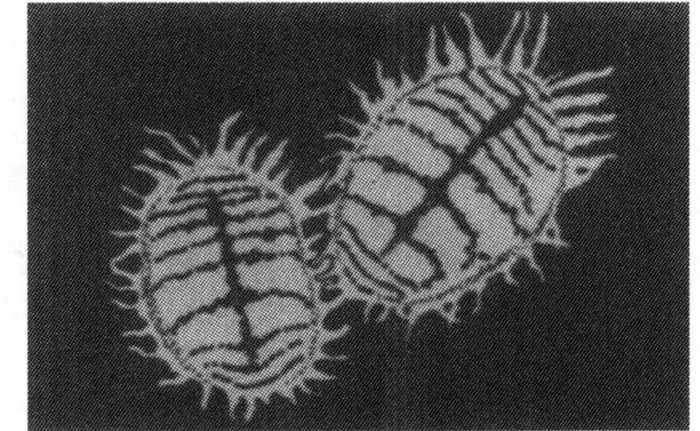
Mealybugs

The programme now includes the training of African crop protectionists and cassava farmers through extension. It also includes the establishment of national biological control programmes across the African continent.

Contact: Dr Hans R. Herren, Director, Biological Control Program, International Institute of Tropical Agriculture (IITA), Benin Research Station, P.O.Box 08-0932, Cotonou, Republic of BENIN, telephone +229-360188/301994, telex 5329 ITABEN, telefax +229-301466.

SLOPING AGRICULTURAL LAND TECHNOLOGY

Mr Harold R. Watson
MBRLC
Davao City
PHILIPPINES

SALT is a complete agro-forestry system for sustainable cultivation of sloping land in humid tropical countries.

Increasing population pressure has forced landless farmers to clear and cultivate sloping forest lands. After a few years the land is wasted and will not recover within a reasonable time. This happens e.g. in the Philippines, but also in other hilly and humid tropical counties.

Mr Watson has developed a complete agricultural - or agro-forestry - system, which he has called 'Sloping Agricultural Land Technology' or SALT.

The essence of SALT is to plant hedges of nitrogen-fixing trees or shrubs along the contour of sloping land. Annual food crops or fruit trees are planted in the strips between the hedges.

The trees or shrubs stabilize the soil of the sloping land with their root systems and reduce soil erosion by acting as barriers to runoff. By fixing nitrogen from the air they produce leaves, which provide protein rich animal fodder as well as nitrogen rich green manure. They also provide fuel wood, fencepoles and other materials for domestic use.

SALT is labour-intensive and at the same time high-yielding and sustainable. It has made it possible for farmers to form permanent settlements and to build real homes.

The complete programme includes the training of local farmers and a variety of forms of help to get started.

Contact: Mr Harold R. Watson, Director, Mindanao Baptist Rural Life Center (MBRLC), PO Box 94, Davao City, PHILIPPINES.

PALM OIL EXTRACTOR

Mr Carl Bielenberg
ATI
Washington, DC, USA

A manually operated press for the extraction of vegetable oil from palm fruits.

This press for palm fruit is of the continuous expeller screw type. The shaft is vertical and is turned manually and directly by a capstan, thus eliminating the need for costly gears. The press depulps and presses the palm nuts in the same operation, producing an extraction rate of 16% from *Tenera* fruit.

This manual expeller press is currently being produced in Cameroon, and was scheduled to be manufactured in Zaire in the beginning of 1990. As the design has eliminated the need for imported gear reducers, the manufacturing costs have been lowered from about 3,500 to about 1,500 USD. The same press is found in operation also in Congo, the Central African Republic and Guinea-Conakry.

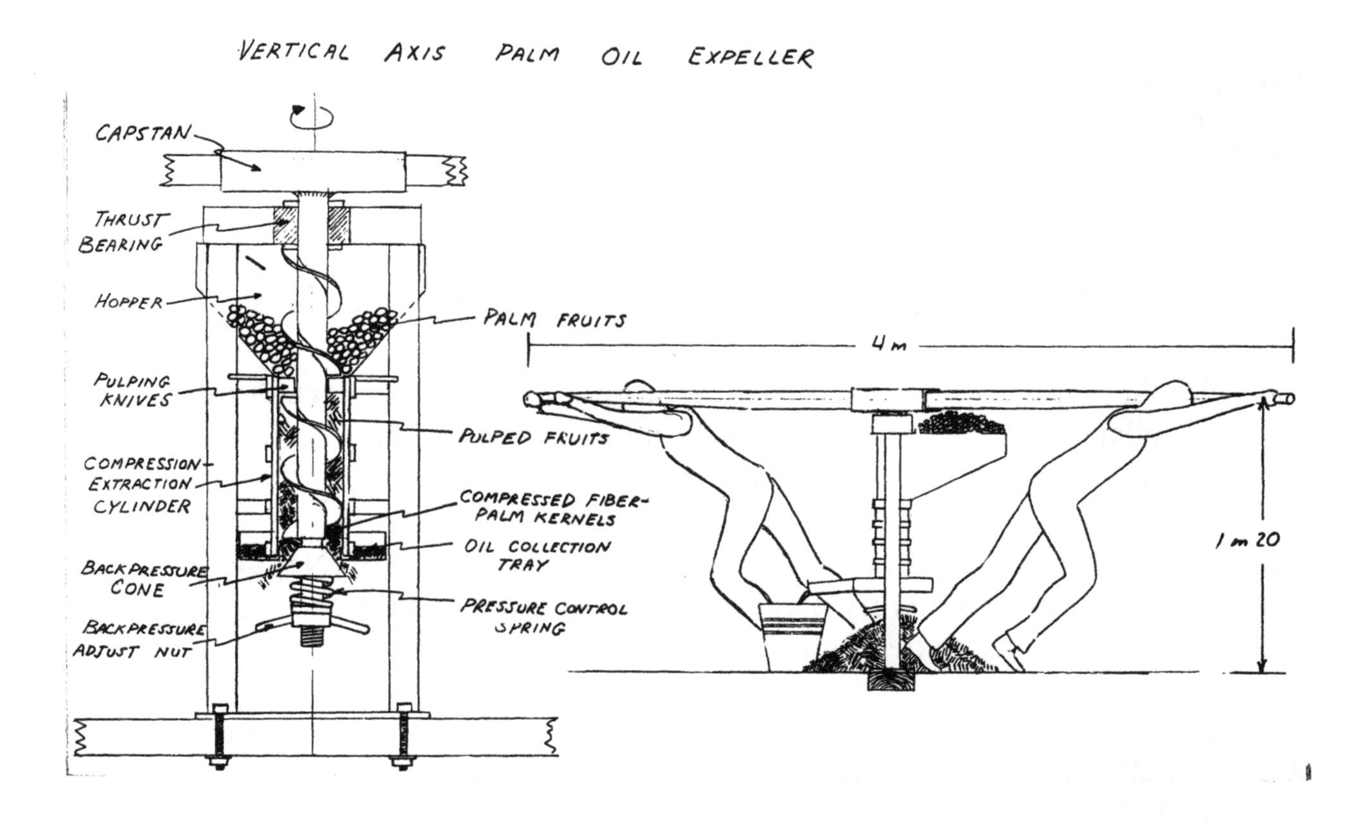

Contact: Mr Carl Bielenberg, c/o Appropriate Technology International (ATI), 1331 H Street NW, Washington, D.C. 20005, USA, telephone +1-202-879-2900, telefax +1-202-628-4622, telex 64661 ATI.

FARMERS ASSOCIATION

Mr R. L. Gapare
NFAZ
Box 3087, Harare
ZIMBABWE

Systematic organization of help to small-holder farmers with loans, advice, grain, transport etc.

The National Farmers Association of Zimbabwe (NFAZ) has initiated a number of schemes to help small farmers boost their productivity and incomes. The following are only a few examples:

- Organization and co-ordination of rural transport services.
- Help to gain access to grain bags on credit.
- Training through the organization of field days, agricultural shows and regular consultative meetings.

As a result of all these activities, the average smallholder's maize, cotton and sunflower yields have doubled.

Contact: Mr Robinson Lyisias Gapare, President, National Farmers Association of Zimbabwe (NFAZ), P.O.Box 3087, Harare, ZIMBABWE, telephone +263-4-737733/4, telex 26217 VARIMI ZW.

VEGETABLE OIL RECOVERY

Dr Felix A. Ryan
Ryan Foundation
Madras, INDIA

Two simple devices for crushing of oil seeds, and for steam distillation of essential oils.

The *Chiuri* is a simple animal-drawn oil seed crusher from Nepal, used for grinding nuts, seeds etc. Every part is made of wood, including the mortar, which is carved out of one large tree trunk.

The distillation unit is a simple, locally made device for steam distillation of essential and non-edible oils. It is used for the recovery of oil from all kinds of raw materials such as nuts, bark, seeds, leaves or roots.

Contact: Dr Felix Augustine Ryan, Ryan Foundation, 8 West Mada Street, Srinagar Colony, Saidapet, Madras 600 015, INDIA, telephone +91-44-411993.

FARM SCIENCE CENTRES

Dr C. Prasad
Krishi Bhavan
New Delhi
INDIA

Krishi Vigyan Kendras (KVK) is a social innovation aiming at the education and agricultural training of the poorer rural population.

The concept of Krishi Vigyan Kendra (KVK) comprises vocational training of farmers, both men and women, as well as farm youths and school drop-outs. It also includes the training of field-level extension functionaries and first-line demonstration of the latest appropriate agricultural technologies, relevant to the respective district. The basic principle is 'teaching by doing'. The poorest of the poor have the priority in the KVKs.

Contact: Dr C. Prasad, Deputy Director General, (Agricultural Extension), Indian Council of Agricultural Research, Krishi Anusandhan Bhavan, Dr K.S. Krishnan Marg, Pusa, New Delhi 110 012, INDIA, telephone +91-11-5731277, telex 031-62249 ICAR IN.

WINTER TRITICALE

Dr Tadeusz Wolski
Danko Plant Breeders
Warsaw, POLAND

Development of new Triticale varieties, which are suitable for poor, acidic soils and harsh winter climates.

Triticale is a hybrid between wheat (*Triticum*) and rye (*Secale*), which can grow well and give high yields on problem soils, such as acidic ones. It gives higher yields in relation to rye under most conditions and to wheat under marginal growing conditions. It requires less nitrogen fertilizer and pesticides. It has good flour and nutritional qualities, very similar to the wheat parent.

This plant will add more value to the crop and has brought an important option to farmers who must cope with land of marginal quality or with harsh winter conditions.

Contact: Dr Tadeusz Wolski, Scientific Director, Danko Plant Breeders, Danko, ul Wspólna 30, 00-930 Warsaw, POLAND, telephone +48-2-210311 x 264, telex 814-288 PL.

LACTIC ACID FERMENTATION FOR FOOD PRESERVATION

Professor S. Mukerjee
Jadavpur University
Calcutta, INDIA

A process for lactic fermentation of vegetables or green fruit as a safe, low-cost method of preserving them as food.

Professor Mukerjee has developed a process for lactic acid fermentation as a means of preserving surplus vegetables and ripe, as well as green fruit, for food at the farm level.

The fermenter can be extremely simple e.g. an ordinary plastic pail or barrel. Thus the investment is very low and the only inputs are vegetables, labour and salt (NaCl). No heat or other form of energy is required.

Contact: Professor Sunit Mukerjee, Department of Bioscience and Bioengineering, Jadavpur University, Calcutta, INDIA.

CROP IMPROVEMENTS

Prof. F. Perez Lopez
Universidad de Trujillo
PERU

Improvement of traditional crops by introducing new genes, for better yields, in the seed material.

Many crop plants, such as wheat, kiwicha and asparagus, have been improved and made more suitable for local growing conditions. As a result new types of crops have been introduced and the yield of traditional crops has been much increased e.g through a shortening of the production cycle.

Professor Perez Lopez has collected local 'plant genes', representing good resistance against diseases and pests, or better productivity under local climatic conditions. These genes have been crossed into traditional crop seeds, resulting in better properties than in the imported seed material, which has now been replaced. The present list of improved crops comprise: Wheat, Qinua, Asparagus, Barley, Kiwicha, Cotton, Beans and Maize.

Contact: Professor Felipe Perez Lopez, Universidad National de Trujillo, PERU, telephone +51-44-242122 or +51-44-248583.

SWILL-MILL FOR ANIMAL FEED

Mr Jorge del Rio
SRI
La Habana
CUBA

An integrated system for the production of animal feed from waste from agriculture, animal processing industry and large kitchens.

Waste from agriculture, agro-industries, food industries, fisheries, large kitchens, city garbage, dead farm animals etc, is processed into a wet, sterile feed product. The resulting paste can be preserved and stabilized by mixing with molasses, which also means an enrichment with nutrients. The net effect of the process is a reduced water content in the final product.

Contact: Mr Jorge del Rio Navarrete, Head, Engineering Department, Swine Research Institute (SRI), Gaveta Postal No.1, Punta Brava, La Lisa, Ciudad de la Habana 19200, CUBA.

MINI GRAIN MILL

Mr R. Shankara
Dr N.G. Malleshi
Dr H.S.R. Desikachar
CFTRI, Mysore
INDIA

A grain milling process, consisting of moistening the grain with water in order to reduce the content of bran in the flour or semolina.

This is a complete small-scale milling process for village use. The key operation is the incipient moistening of the grain with 1.5 to 3% of water and tempering for about 5 minutes, in order to toughen the bran. This conditioning makes it possible to use a low-cost plate grinder with good results. After sifting and aspiration the resulting flour or semolina is largely free from husk and bran.

Contact: The Director, Central Food Technological Research Institute (CFTRI), Attn: Mr R. Shankara, Scientist, Cereal Science and Technology Discipline, Mysore 570 013, Karnataka, INDIA, telephone +91-821-37253, telex 846-241 FTRI IN.

MODIFIED CLAY AGAINST INSECTS

Prof. S.K. Majumder
Prof. J.S. Venugopal
CFTRI
Mysore, INDIA

A method of producing an insecticide from ordinary clay. It is used in dry long-term storage of food grains and seeds.

This is a novel pesticide, which uses ordinary clay, i.e. minerals such as *montmorillonite, illite* or *kaolinite*, as raw material.

In many countries the storage of dry food grains and seeds between the seasons is a great problem, because of infestation by insects. The inventors claim that they have developed a derivative of clay, which kills the insect larvae, but which is harmless for man to consume.

Contact: Professor S.K. Majumder, Additional Director (Retired), Central Food Technological Research Institute (CFTRI), Mysore 570 013, INDIA, telephone +91-821-23046, telex 846-241 FTRI IN.

DROUGHT-RESISTANT ORANGE PLANTATIONS

Mr A.P. Kinkhede
Pipla Kinkhede, Nagpur
Maharashtra, INDIA

The planting of *Acacia* shrubs between orange trees in order to improve the drought resistance of the plantation.

By planting a fast growing Acacia shrub - locally known as *Subabul* - between the orange trees, the moisture content of the soil and the moisture regime for the entire orange crop has been upgraded. The plantation's ability to survive drought conditions has been much improved, which has been shown in comparison with traditional plantations in the same area. The activities have expanded to cover many forms of reforestation and education. The integrated food crop and forestry programme has improved the general availability of water, food, fodder and firewood for a population of 1700 people and 500 head of cattle.

Contact: Mr Ajay Prabhakar Kinkhede, Shri Ganesh, 46 Khare Town, Dharampeth, Nagpur-440010, Maharashtra, INDIA, telephone +91-712-531684.

MUSHROOMS FROM RICE STRAW

Dr H.S. Garcha
Punjab Agric. Univ.
Ludhiana
INDIA

A system for producing mushrooms on a substrate consisting of composted rice straw. Spent substrate is recirculated as animal feed or mulch.

A system for the cultivation of mushrooms of the genera *Agaricus, Pleurotus* and *Volvariella,* on a substrate derived mainly from agricultural solid waste. The raw material is usually paddy (rice) straw, that is less than one year old. The moistened straw is heaped in the form of stacks, without any additives. After about 84 hours the cores of the stacks are ready for use as substrate for mushroom cultivation. The harvested mushrooms are preserved by sun-drying to about 6% moisture. The spent substrate has a potential value as feed to animals or as mulch for soil conditioning.

Contact: Dr Harnek Singh Garcha, Department of Microbiology, Punjab Agricultural University, Ludhiana, Punjab, INDIA.

IMPROVEMENTS IN AFRICAN FARMING SYSTEMS

Dr Bede N. Okigbo
United Nations Univ.
New York City, USA

Improvements in African farming systems and in nutrition, by effective use of wild indigenous African crops.

- Evaluation of the nutritional value of the proteins in the African yam bean (*Sphenostylis stenocarpa*), and its use for treatment of kwashior.
- Cultivation of the winged bean (*Psophocarpus palustris*) for live mulch, used as nitrogen fertilizer and soil conditioner.
- Mulching systems, based on crop residues, which can double the yield of e.g. cassava and yam productions.

Contact: Dr Bede N. Okigbo, Director, United Nations University, Office in North America, United Nations, Room DC2-1462-70, New York, New York 10017, USA, telephone +1-212-963-6387, telefax +1-212-371-9454, telex 422311 UN UI

COFFEE PLANTS RESISTANT TO DISEASES AND PESTS

Dr Alcides Carvalho
Instituto Agronômico
Campinas, BRAZIL

Development of basic genetics and new lines of coffee resistant to a number of diseases and pests.

Development and breeding of new lines of the coffee plant *Coffea arabica*. Of particular value were the breeds which are resistant to the leaf-rust disease (*Hemileia vastatrix*), nematodes (*Meloidogyne* sp.) and the leaf miner (*Perileucoptera coffeella*). Nowadays Brazilian farmers are provided with lines giving good yields and being naturally resistant to diseases and pests, thus avoiding the use of costly and polluting pesticides. A number of other difficult-to-study crops are now being researched.

Contact: Dr Alcides Carvalho, Seção de Genética, Instituto Agronômico, Caixa Postal 28, 13100 Campinas, SP, BRAZIL, telephone +55-192-410511, telex 55191059.

HERBA-FARMING

Dr Pulak Roy
Calcutta
West Bengal, INDIA

A system for farming, aiming at optimal health rather than maximum volume of production.

The project will be implemented in six steps:

- Creating awareness among villagers about the concept of farming as a tool for health.
- Re-establish faith in the efficacy of traditional vegetables, tubers and herbs in preventing diseases.
- Educating villagers in identifying useful herbs.
- Arranging demonstration plots.
- Distributing plants to the promoters of the project.
- Further planting of herb gardens in the villages until the need is satisfied.

Contact: Dr Pulak Roy, 42/113 East End Park, Calcutta, 700 039, INDIA, telephone +91-33-40-9002.

FISHING AND AQUACULTURE

This target area received a rather large number of interesting nominations. We obtained contributions from all continents, but the majority of them from Asian countries. Within the area of fishery there was a great variety of innovations, ranging from small scale enterprises, to advanced micro-computer software, to be used as a tool for management and conservation of tropical multi-species fish stock. The fishery nominees have worked both in coastal areas and in freshwater reservoirs.

Fish aggregating devices were described in two nominations. In clear tropical waters, with smooth sand bottoms, the fish are normally dispersed over wide areas, which make them difficult to catch. By using various kinds of fish aggregating devices, the fish become more concentrated.

New types of fishing crafts are essential for coastal fishermen in tropical areas, especially where it is difficult to build ports. Two types of crafts were nominated, and one of these innovations was given an honorary award. It consists of a beach landing craft, which can start from, and land on, shallow sandy beaches exposed to heavy swell. This innovation has made it possible for fishermen to operate more than one day at a time and to reach fish populations further out at sea, and thus compete with the international deep sea fishing fleets.

Both freshwater and marine forms of aquaculture were represented. All these innovations aim at a sustained yield. They covered techniques for small scale fish-farming, but also advanced methods for larger enterprises. One innovation minimises the water volume in fish production tanks in arid areas, and consists of a simplified closed system with recirculation of the water through bio-filters.

The nominations were dominated by methods of reproduction and cultivation of finfish, shellfish and seaweeds. Many nominees have developed techniques for artificial spawning of fish and shellfish, especially of fresh water fish such as carp, mullet and pike-perch. The IDEA award winner has developed hatchery methods for rural fish farming, including the induction of spawning, hatching of eggs, rearing of fry and fingerlings as well as feeding with agricultural waste products.

Hans Ackefors

INTEGRATED FISH FARMING

PRIZE WINNER

Dr E. Woynarovich
Budapest
HUNGARY

New techniques for the management of freshwater finfish farming, integrated with animal husbandry.

Professor Woynarovich has developed a number of techniques related to artificial propagation of freshwater finfish and their farming. The following are only a number of examples:

- Routine methods for hypophysation, or use of synthetic gonadotrophin analogues, in *Cyprinids* and *Characids* in order to induce ovulation and spermiation.
- Treatment of common carp (*Cyprinus carpio*) eggs with urea and common salt, followed by a short dip in a tannin solution, in order to improve the yield in funnel type incubators.
- Maintenance of natural populations of pike-perch (*Stizostedion lucioperca*) through yearly stocking with hatchery larvae. Wild fish are induced to spawn in artificial nests, and the eggs are hatched in spray-chamber incubators.
- The Carbon Manuring Technique (CMT) for maintaining the natural productivity of fish-ponds, by adding fresh manure from pigs or cattle. CMT includes methods of dissolving and dispersing the manure.

Sorting out the Tambaqui (*Colossoma macropomum)* females for hypophysation in Pirassununga, Brazil.

Contact: Dr Elek Woynarovich, Attila ut 121, H-1012 Budapest, HUNGARY, telephone +36-1-1753-418, telefax +36-1-1753-418.

BLC-TYPE FISHING VESSEL

HONORARY AWARD WINNERS

Mr Ø. Gulbrandsen
Grimstad, NORWAY
Mr G. Pajot
Mr R. Ravikumar
Madras, INDIA

A beach-landing craft and fishing vessel, provided with a completely retractable propulsion unit, making it possible to fish far out at sea from a shallow coast.

This team has developed a new kind of fishing vessel, which can be used on tropical beaches and shores, that lack proper fishing ports. It is a type of beach-landing craft (BLC), which can start and land from a shallow and exposed coast.

The new fishing vessel has been named IND-20 and is a motorized BLC that can negotiate heavy surfs. It is made of fibreglass reinforced plastic laminate, stiffened with transverse plywood frames. The hull is slender and gives little resistance in the water. The bottom is without a keel. It is driven by a small diesel engine, fitted together with the rudder in a separate watertight and retractable propulsion unit. The engine can be started on the beach before launching and vice versa. IND-20 can also be propelled by sails, in which case the retractable dagger-board is lowered.

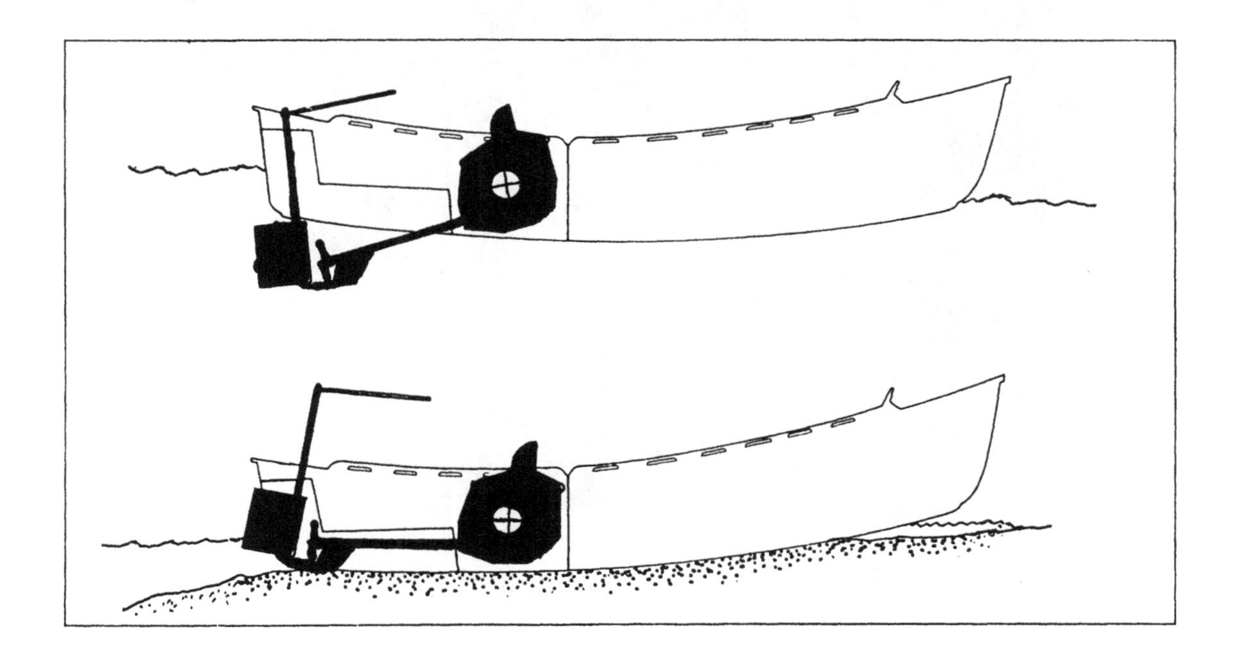

Contact: Bay of Bengal Programme, 91 St Mary´s Road, Postbag 1054, Madras 600 018, INDIA, telefax +91-44-836102, telex 41-8311 BOBP.

BLACK TIGER PRAWN AQUACULTURE

Dr I-Chiu Liao
TFRI
Keelung TAIWAN

A complete system for Black Tiger Prawn cultivation, including techniques for larval rearing and growout.

Dr Liao has developed reliable larval rearing and growout techniques for the black tiger prawn (*Penaeus monodon*), also called the Grass Prawn. He has also worked with other species of *Penaeid* prawns. TFRI has characterized the nutritional requirements for each larval stage, and has developed appropriate feed formulations for each of them.

Dr Liao has also initiated the design of hatcheries as well as growout and management systems. His work has closed the gap between the hatchery and the growout stage of prawn culture, and has combined it all into a complete system for artificial propagation. Prawn culture is now independent of wild seed-stock, which previously meant considerable fluctuations in numbers from year to year.

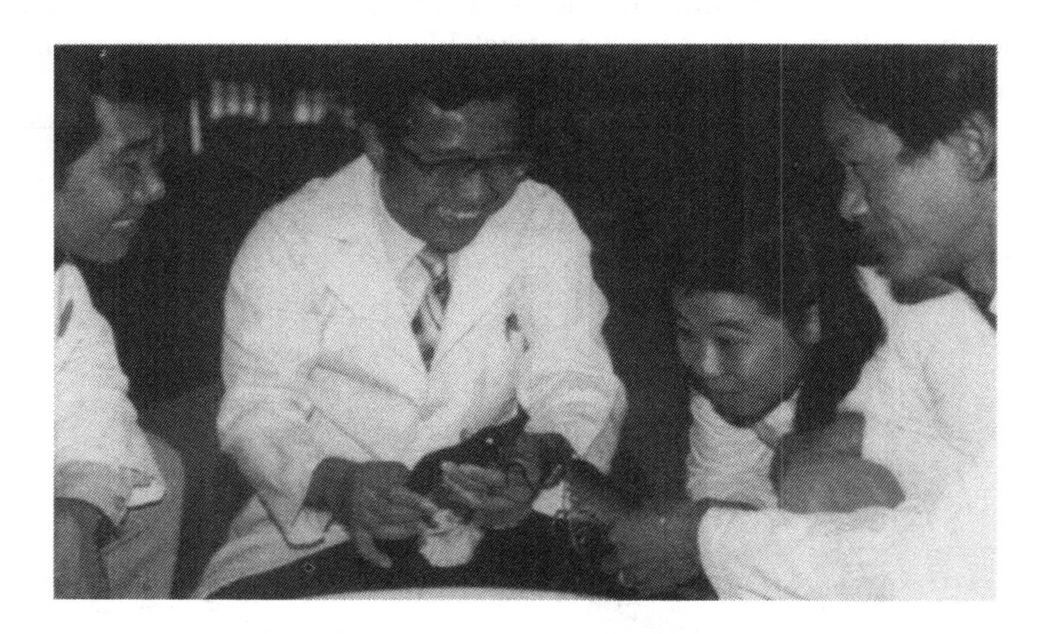

A lively discussion over black tiger prawns

Contact: Dr I-Chiu Liao, Director General, Taiwan Fisheries Research Institute (TFRI), 199 Hou-Ih Road, Keelung 20220, TAIWAN, telephone +886-2-4628283, telefax +886-2-4629388.

ARTIFICIAL SPAWNING OF CARP

Dr Wang You-Lan
Academia Sinica
Shanghai, CHINA

A system for inducing spawning in various carp species by injecting hormones.

The teamwork behind this system involves also Dr Wang Ying-Tian, Mr Lin Zhi-Chun, Mr Xu Guo-Jiang, Mr Pong Shi-Yi, Mr Liu Shi-Fan, Mr Dai Rong-Xi and Mr Zeng Mi-Bai.

A system for improved pond culture, artificially induced spawning of carp in aquaculture, egg collection and fry hatching has been developed. This has eliminated the necessity of obtaining fry from natural waters. This reduces the dependence on a fluctuating supply of wild fry.

Mature carp, both females and males, are injected with human chorionic gonadotrophin (HCG) hormone and are kept in running water. This results in an induced spawning and fertilization of the eggs. The hatching is carried out in circular tanks with running water. Proper timing is the key to success. Normal development and hatching of carp eggs depends on fertilization of oocytes at an optimal stage of maturation. Silver carp (*Hypophthalmichthys molitrix*) for example, offers a period as short as 2 hours at 25 degrees centigrade for spawning and optimal fertilization. This period comes about 10 to 14 hours after the hormone injection.

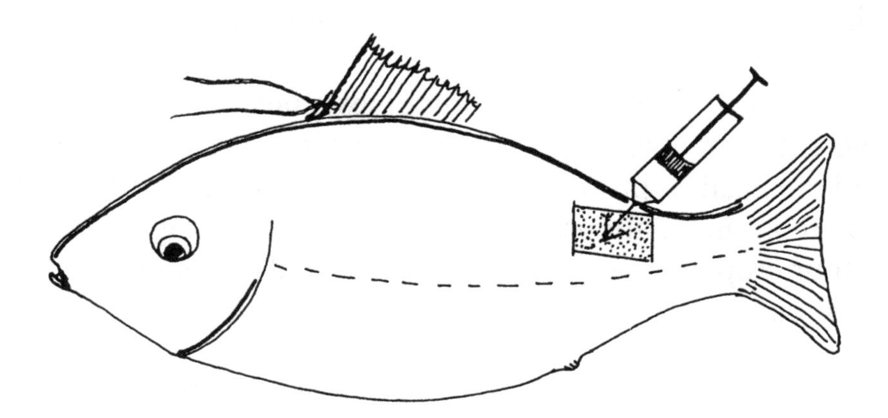

Contact: Professor Wang You-lan, Shanghai Institute of Cell Biology, Academia Sinica, 320 Yo-Yang Road, Shanghai 200 031, CHINA, telephone +86-21-4331090, telefax +86-21-4331090.

SEAWEED CULTIVATION

Professor Maxwell Doty
University of Hawaii
Honolulu, USA

A complete system for marine cultivation of red seaweeds in small production units in tropical coastal areas.

A system for growing red seaweeds of the genus *Eucheuma* in the coastal zone of the sea. These seaweeds are of great economic importance for many tropical countries, as they are the raw material for the production of hydro-colloids, such as carrageenans, which have applications in the food industry.

More than two decades ago the leading companies of the world, which were involved in the production of industrial colloids from tropical seaweeds, and especially carrageenan from the red algae *Eucheuma*, realized that the world demand for colloids was far greater than the wild crops of seaweeds could supply. Signs of depletion of the wild stocks through over-harvesting were beginning to be noticed in the Philippines.

Professor Maxwell, in co-operation with Marine Colloids Inc. USA and with Dr G.T. Kraft, Dr R.E. Wreede, Dr K.E. Mshigeni and Mr V. Alvarez, embarked on a series of field experiments and pilot farming operations in order to develop methods for producing commercial crops of *Eucheuma* through marine agronomy.

Professor Doty's innovation made use of scientific information on the field distribution ecology and growth dynamics of *Eucheuma* species, with special reference to those occurring in the marine waters of the Philippines.

The technology, as developed for the Philippines, consists of growing the seaweeds on nylon mono-filament lines, supported by mangrove stakes. Vegetative fragments of the seaweed are used as propagating material. A farm of this kind can support about 40 to 100 thousand thalli per hectare. With good management it is possible to harvest up to four crops per year. Harvested seaweeds are sun-dried by the farmers and sold to local buyers.

This seaweed mariculture is carried on by individual families in the coastal village communities, thus improving the socio-economic level of the whole area.

Contact: Professor Emeritus Maxwell S. Doty, Botany Department, University of Hawaii, Honolulu, Hawaii 96822, USA, telephone and telefax +1-808-9568561.

INTENSIVE BABY FISH FARMING

Dr S. Appelbaum
Ben-Gurion University
ISRAEL

An intensive recirculating system for mass rearing of fish larvae. The system is independent of climatic factors.

Dr Appelbaum has developed a simplified closed system for the mass rearing of fish larvae, utilizing only a small volume of water. It is based on the reuse of water that has been recycled through a conventional biological water treatment process, making it reasonably independent of climatic conditions. In this system, fish larvae are fed with manufactured dry feed. This makes it possible to produce baby fish the year round, disregarding the availability of natural feed.

A fish farm consists of two main units:

- The rearing unit.
- The water purification unit.

The rearing unit has two distinct features, both of which are designed to facilitate the removal of detritus and waste:

- Walls and bottoms have such angles as to make possible cleaning by simple flushing.
- A water outlet provided with a low voltage electrode 'gate', that prevents escape of fish, but lets particulate organic material pass.

The water purification unit has two main components:

- A mechanical filter for the removal of solid particulate waste.
- A biological gravel-bed filter for the removal of soluble organic compounds.

The system offers a high rate of fish larvae production, intensive water use, control of diseases and parasites, and full control of water quality parameters.

Contact: Dr Samuel Appelbaum, Aquaculture Unit, The Jacob Blaustein Institute for Desert Research, Ben-Gurion University of the Negev, Sede Boker Campus, ISRAEL 84990, telephone +972-57-565835, telefax +972-57-555058, telex 5280 DIRBG IL.

RAFT FOR COASTAL FISHING

Dr C.V. Seshadri
MCRC, Tharamani
Madras
INDIA

An improved version of the traditional Kattumaram, in which the wooden logs have been replaced by sealed plastic pipes.

This invention is an improved version of the traditional *Kattumaram* (Catamaran), which is used for coastal fishing and many other purposes on the east coast of India. In the new version, the wooden logs have been replaced by sealed plastic pipes.

Dr Seshadri has developed the design, as well as the techniques required for manufacturing in the new material of construction. The raw material is pipes made of high density polyethylene (HDPE) with an outer diameter of 160mm. This particular plastic can be made from alcohol, derived from native sugar cane, as the only raw material. The pipes are sealed and shaped and six of them are united into one fishing raft. The vessel can be manufactured locally with simple tools.

The new design is much lighter and easier to handle on the beach, as it is self-draining. The expected average life-length is 15 years, which is an improvement on the old design.

Rafts of similar design and material of construction can be used as fish aggregation devices, river barges, cargo containers, for shellfish farming, etc.

Contact: Dr C.V. Seshadri, Shri. AMM Murugappa Chettiar Research Centre (MCRC), Tharamani, Madras 600 113, INDIA, telephone +91-44-411937, +91-44-419369, telex MCRC CARE 041 7132 CUMI IN.

OYSTER FARMING

Mr J. Rodríguez
in a team of six
MIP, CUBA

A complete system for oyster farming in the intertidal marine zone of Cuba.

This is the work of a team, consisting of Jorge Rodríguez Montoro, Jose Antonio Frías, Cristina Perera, Carlos Luis Felipe, Rafael Rubio and Antonio Morales.

A system for oyster farming in Cuba, based on the species *Crassostrea rhizophorea*. The technology consists of the following basic steps:

- The cutting and anchoring of collectors.
- First cleaning of the collectors.
- Second cleaning of the collectors and the starting up of the production cycle.
- Maintenance of the collectors in the intertidal zone of the sea.
- Harvesting.

The collectors consist of terminal branches of the red mangrove tree *Rhizophora mangle.*

Oyster farming in Cuba is now concentrated to a few areas, where the natural environment is less influenced by industrial activities. The new development programme has the following components:

- Introduction of new species of bivalves.
- Improved supply of seed (larvae) through induced spawning.
- Expansion of the farming area by using floating collectors.

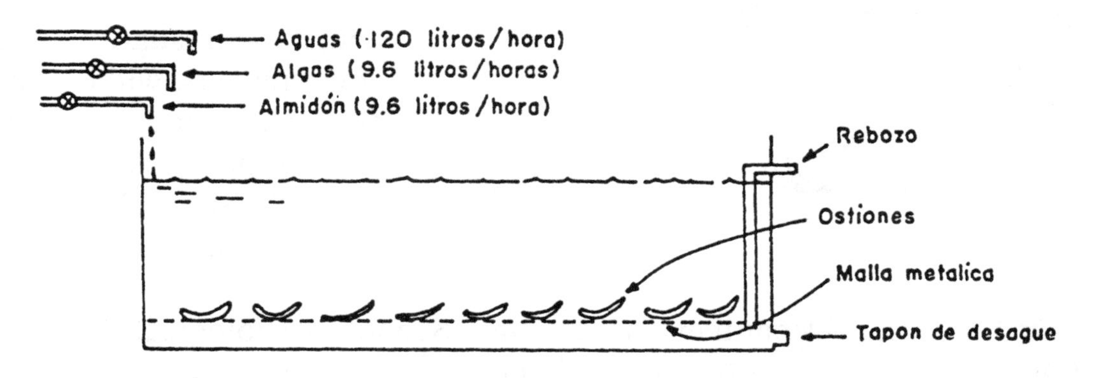

Contact: Mr Jorge Rodríguez Montoro, Centro de Investigationes Pesqueras, Ministerio de la Industria Pesquera de Cuba (MIP), CUBA.

GIANT CLAM AQUACULTURE

Dr J.S. Lucas
James Cook Univ.
Townsville, AUSTRALIA

A complete system for cultivating Giant Clams in Pacific Island Nations.

This innovation is the product of funding by the Australian Centre for International Agricultural Research (ACIAR) and co-operation between James Cook University of North Queensland (JCU), Fiji Fisheries, University of Papua New Guinea and two universities in the Philippines, as well as the Pacific Island Nations of Cook Islands, Kiribati, Tonga and Tuvalu.

The species selected for the project at JCU was *Tridacna gigas*. The technology, however, is applicable to all eight giant clam species. The most important and innovative steps in the new cultivation system are as follows:

- Spawning induction through injection of serotonin into the giant clam gonad.
- Rearing larvae in a flow-through culture.
- Use of micro-encapsulated diets in rearing the larvae.
- Intertidal culture of juvenile clams in ‘lines’, ‘covers’ and ‘exclosures’.

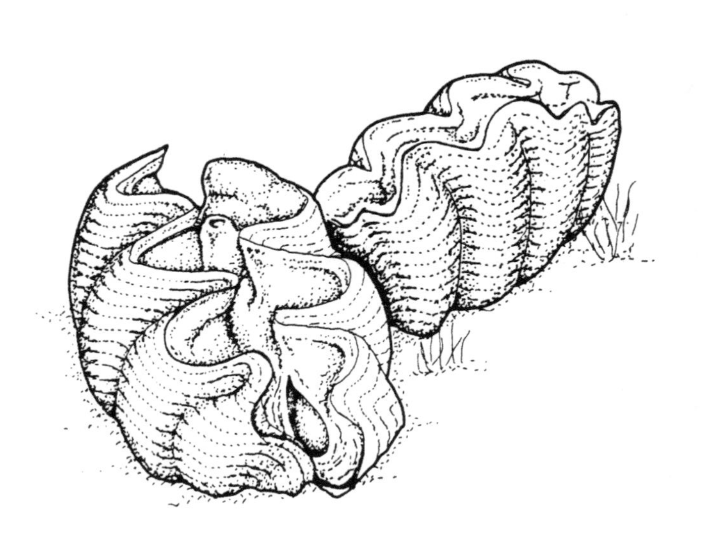

The technology for juvenile clams works well in a fishing village environment. The fisheries departments of the Pacific island nations are capable of rearing larvae. Giant clams have been kept at very high densities in shallow reef environments, without any deleterious effects on the environment. They gain most of their nutrition from symbiotic algae.

Contact: Dr J.S. Lucas, Department of Zoology, School of Biological Sciences, James Cook University of North Queensland (JCU), Townsville, Queensland 4811, AUSTRALIA, telephone +61-77-814412, telefax +61-77-251570, telex AA47009.

SEAWEED FARMING

Prof. K.E. Mshigeni
Dar es Salaam
TANZANIA

A system for small scale marine cultivation of seaweeds, and their processing into phyco-colloids.

Professor Mshigeni has been engaged in research on seaweeds as a potential source of medicinal substances, as fodder for livestock, as agricultural fertilizer, as food for man and as fishbait. With his students he has isolated and characterized economically important colloids from a variety of Western Indian Ocean Seaweeds, and has identified those which have a potential for mariculture, and as raw materials for local phyco-colloid extraction industries.

Professor Mshigeni has successfully operated pilot seaweed farms in Tanzania and has demonstrated their economic feasibility for carrageenan-rich species of the genus *Eucheuma*.In Zanzibar alone, seaweed farming currently provides nearly 2000 families with their livelihood, and many other locations are under development.

Seaweed farming results in increased marine bio-productivity and in a richer stock of wild fish around the farm sites, and thus also in better conditions for the local fishermen.

The next step in the development will be the establishment of local phyco-colloid extraction plants.

Other seaweed genera under study are *Hypnea, Gracilaria, Sargassum, Digenia* and *Chondria*, and other nations involved are Kenya, Mauritius, the Seychelles, Namibia, Qatar, and Mozambique.

Contact: Professor Keto Elitabu Mshigeni, Department of Botany, University of Dar es Salaam, P.O.Box 35091, Dar es Salaam, TANZANIA, telephone +255-51-49192 ext. 2010, telefax +255-51-48274, telex 41327 UNISCIE or 41561 UNIVIP TZ.

MONITORING FISH POPULATIONS

Dr Daniel Pauly
ICLARM
Manila, PHILIPPINES

A computer software package for estimating vital statistics of fish populations from length-frequency measurements.

Dr Pauly has developed a computer programme called ELEFAN (Electronic Length-Frequency Analysis), which allows the estimation of growth and related parameters, such as mortality, gear selection and recruitment, from widely available length-frequency data. It is now used for the assessment of tropical fish populations, and for the rational management of fisheries in many developing countries. ELEFAN is particularly useful for multi-species finfish stocks, constituent species of which are small and shortlived, as well as for invertebrates.

Contact: Dr Daniel Pauly, International Center for Living Aquatic Resources Management (ICLARM), P.O.Box 1501, Makati, Metro Manila 1299, PHILIPPINES, telephone +63-2-818-0466, telefax +63-2-816-3183, telex (ETPI) 64794 ICLARM PN.

MARITIME ENVIRONMENT PROTECTION

Mr G.P. Livanos
HELMEPA
Athens
GREECE

An organization, which aims at reducing marine pollution by training seamen and merchant marine officers in environment protection.

Mr Livanos has been the innovative force behind HELMEPA, the Hellenic Marine Environment Protection Association. This organization is the result of a voluntary commitment from representatives of the Panhellenic Seamen's Federation and the Union of Greek Shipowners. The main objective of HELMEPA is to instill an environmental consciousness through all sectors of the shipping industry.

Contact: Mr Georges P. Livanos, Chairman, Hellenic Marine Environment Protection Association (HELMEPA), 5 Pergamou Street, N. Smyrni, GR-17121 Athens, GREECE, telephone +30-1-9343088/ 9341233, telefax +30-1-9353847, telex 223179 HELM GR.

LOBSTER FISHING

Mr Raúl Cruz
in a team of four
MIP, CUBA

A management system for the monitoring and regulation of spiny lobster fishing in CUBA

This is the work of a team, consisting of Mr Raúl Cruz Izquierdo, María Estela de León González, Rafael Puga Millán and Romárico Sotomayor Parra.

The innovation consists of a management system to optimize the catch of several species of spiny lobster and snapper *(Panulirus argus* and *Lutjanus synagris)* on the ocean shelves around the shores of Cuba.

Contact: Mr Raúl Cruz Izquierdo, Centro de Investigationes Pesqueras, Ministerio de la Industria Pesquera de Cuba (MIP), CUBA.

ARTIFICIAL INSEMINATION OF CARP

Dr Adib Saad
Tishreen University
Lattakia, SYRIA

A system for artificial insemination of carp, including preservation of semen.

Dr Saad has developed a system for artificial insemination of Carp (*Cyprinus carpio*), and in particular a method for short-term preservation and storage of carp semen.

Dr Saad has studied the spermatozoa production in carp during the annual cycle, and the effect of hormonal stimulation on the number of collected spermatozoa, their mobility and fertilizing ability as well as storage properties.

This research has resulted in recommendations to local fish-farmers in warm and temperate underdeveloped areas.

Contact: Dr Adib Saad, Tishreen University, P.O.Box 1408, Lattakia, SYRIA, telephone office +963-41-36401, home +963-41-30766, telex 451084 TIUNIV SY.

FORESTRY

The nominations for the IDEA 1990 award in forestry cover a wide range of innovations, from forest genetics and increase of forest yields by use of improved strains, through forest management, to the harvesting and transport of forest products.

Out of all these very important innovations, the one on the management of natural forests in the semi-dry savannah area of West Africa, was chosen for the 1990 prize. Another innovation, on the genetical improvements of tree species and their dissemination to a number of Developing Countries, was chosen for the honorary award.

John Heermans, USA and Sani Sidi, Niger received the prize for their work with the Guesselbodi forest in Niger. They have developed a management system for the 5.000 ha forest, which today is implemented by the farmers living around the forest. It consists of controlled cutting of fuel wood, for the farmers' own use, and for sale in the nearby Niamey. Suitable parts of the forest area are used for grazing. Part of the revenue is used by the farmers for improving wood production and soil conditions in the forest. All measures are organized in such a way that a sustained, or even increased production from the area is maintained.

Besides yielding wood and fodder, natural forests give valuable byproducts, such as edible fruits, fibres for ropes and pharmaceutical ingredients. They also form a valuable gene pool for further development of forestry and agriculture in the tropics. Natural forestry management is very adaptable and has many advantages over the plantation system. It deserves to be much more widely applied in the Developing Countries.

Helmuth Axel von Barner and Henrik Keiding, Denmark, received the Honorary Award for the creation and management of the Danish Seed Centre. This institution has spearheaded and co-ordinated a continuously growing programme that provided training, consulting and other services, supported by. research and development in the fields of genetic improvement, gene resource conservation and seed procurement. The Seed Centre has a reputation for providing technically sound advice and practical support through its participation and leadership in many international projects.

Jöran Fries

NATURAL FOREST MANAGEMENT

PRIZE WINNERS

Mr John Heermans
Mr Sani Sidi
USA and NIGER

A management system for natural savannah forests in semi-arid regions of West Africa

The prize-winners have developed a novel management system for natural savannah forests in semi-arid regions of West Africa. This is an alternative to clear-cutting the forests in order to replant with foreign species of trees.

In the Guesselbodi model management system, the forest was divided into work parcels which were treated successively and underwent controlled cutting and grazing rotations. Research related to the management plan was carried out concurrently. Selected parts of the forest were completely protected from grazing for a number of years, and guards impounded animals that illegally enter the woods. This protection is sufficient to allow natural grasses to recover, which increases infiltration and reduces erosion from heavy monsoon rains, resulting in benefits for the water table.

Local villagers have been involved from the very beginning in the actual work as well as in the decision-making process. They are the people most dependent on the forest´s productivity. They cut fuelwood for their own households, but also for sale. Selected parts of the forest are allocated for cattle-grazing. A certain part of the revenue is used for improving wood production and soil conditions as well as water conservation, in such a way that the system becomes self-reliant and sustainable. Besides firewood, the forest also yields other valuable products such as honey, edible fruits, fibres and natural medicines.

After several years of participation, the villagers agreed to form a producers' co-operative for the management of the forest resources. This has inspired the Government of Niger to grant exclusive legal rights to villages in forest vicinities.

Contact: Mr Sani Sidi, Forest Agent, Forest Land Use Planning Project (FLUP), Ministry of Hydrology and Environment, B.P. 12.520, Niamey, NIGER, telephone +227-722087, or Mr John Heermans, Forestry consultant, 9 Jackson Street, Essex Junction, Vermont 05452, USA, telephone +1-802-879-7946.

TREE SEED CENTRE

HONORARY AWARD WINNERS

Mr H.A. von Barner
Mr H. Keiding
DENMARK

An international seed centre for forest species, supporting sustainable tropical silviculture.

The Tree Seed Centre carries out the collection and handling of seeds, introduction and evaluation of foreign species, and mass production of improved seeds. The Centre is now able to deliver e.g. seeds of teak (*Tectona grandis*), and of two pine species. Other activities include the introduction of bud-grafting and tree climbing methods.

From instruction book on tree climbing for seed collection.

The Centre provides training, consultation and other services, within the fields of forest species genetics, gene resource conservation and seed procurement.

An essential part of the Centre's objectives is to establish national seed centres, through the transfer and exchange of technology and expertise in seed procurement, tree improvement and gene-resource preservation. At present the Seed Centre is involved in national tree seed projects in Thailand, Tanzania, Nicaragua, Sudan and Nepal. The Centre is now being drawn into a wider network, which includes arid and semi-arid areas of Africa.

Contact: Mr Henrik Keiding, DANIDA Forest Seed Centre, Krogerupvej 3A, DK-3050 Humlebaek, DENMARK, telephone +45-42-190500, telex 16600 FOTEX DK, telefax +45-49-160258, or Mr Helmuth Axel von Barner, Plantagevej 5A, DK-3100 Hornbaek, DENMARK, telephone +45-42-201572.

CLONAL SILVICULTURE

Professor B. Martin
ENGREF
Nancy, FRANCE

A clonal silviculture, using Eucalyptus hybrids for planting and reforestation of savannah land in tropical countries.

Dr Martin has developed a system for clonal silviculture of hybrids within the Australian genus *Eucalyptus*. It has been applied in many tropical countries, e.g. for reforestation in the Congo.

His achievements are the result of a sophisticated strategy, which is based on 13 species and 38 provenances. He has been using efficient breeding methods such as grafting, natural and artificial pollination, vegetative propagation, clonal tests, etc.

Eucalyptus hybrid plantations in the Congo have achieved a rotation age of as little as six to seven years on previously deforested land. Thus they will protect these areas from further degradation, at the same time as the wood production lessens the pressure on the natural forests in the area.

The continuous input into the system is limited to seedlings, which are produced in nurseries, requiring only plastic bags, fertilizers and some simple tools.

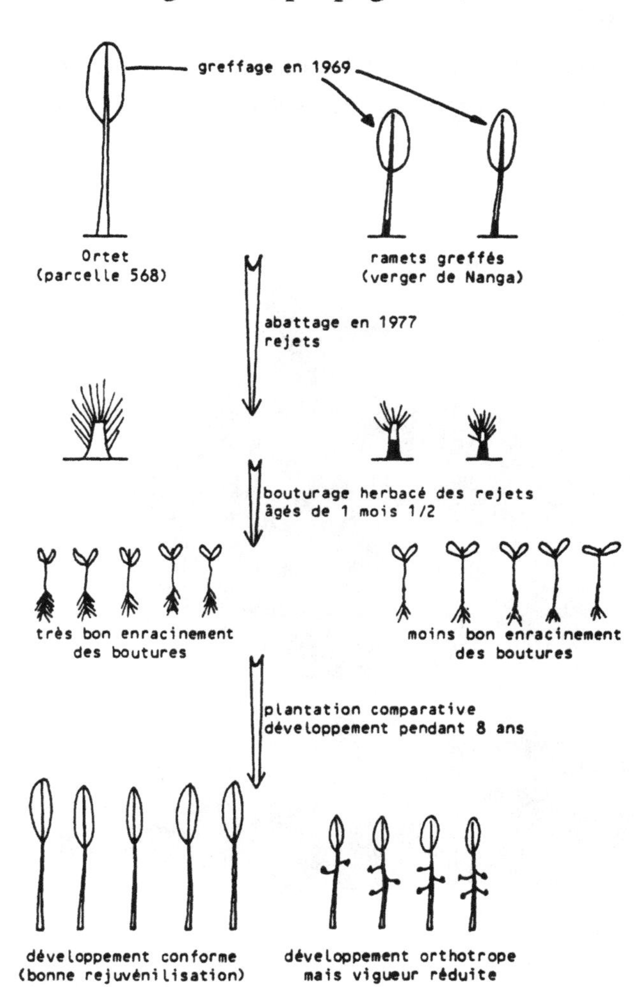

Contact: Professor Bernard Martin, Ecole National du Génie Rural, des Eaux et des Forêts (ENGREF), Centre de Nancy, FRANCE.

CAROB CULTIVATION

Dr N. Ibáñez-Herrera
Trujillo, PERU

A project plan for the cultivation on a large scale of the Peruvian Carob tree.

Mr Ibáñez-Herrera is working with the Peruvian Carob tree (*Prosopis pallida*), which grows wild in the departments of Piura and Lambayeque in the northern tropical, yet arid, coastal parts of Peru.

This Carob tree produces a very strong wood, which is used now mainly for the production of charcoal. It belongs to the pea family of plants (*Leguminosae*) and is able to carry out biological nitrogen fixation (BNF) from the air. The leaves can be used as feed for e.g. sheep. It produces plenty of flowers and is thus a source of honey for local apiculture. The bean-like pod is the raw material for 'algarrobina', a dark viscous liquid with food uses similar to that of carob powder, produced in semi-arid Mediterranean areas from locust beans (*Ceratonia siliqua*).

The Peruvian Carob is relatively resistant against native insect pests, and thus worth cultivating on a large scale.

This project, which is in a very early stage of development, will involve also genes and grafts from the related species *Prosopis haslerii* from Argentina as well as *Prosopis tamarugo*. It also involves inoculation with species of *Rhizobium* in order to improve the BNF, as well as proper water management.

Contact: Dr Nicanor Ibáñez-Herrera, Facultad de Ciencias Biologicas, Universidad Nacional de Trujillo, San Martin 380, Apdo. 315, Trujillo, PERU, telephone +51-44-235841. Private, Paraguay 160, Urbanización El Recreo, Trujillo.

LOG TRANSPORT SULKIES

Mr Erling Fosser
Halden
NORWAY

Two manual sulkies for pulling down hung-up trees, and for the transport of logs over short distances.

Mr Fosser has developed two types of manual logging sulkies, which are particularly adapted to use in developing countries.

The hung-up sulky is used in thinnings to pull down trees, which do not fall after cutting, but get stuck in the branches of adjacent trees.

The skidding sulky (picture) is used for short distance transport of small-sized logs, especially downhill to the nearest roadside. The load can consist of a single log, or bundles, and can be as high as 200 kilogrammes.

Both sulkies are especially useful for management of softwood plantations, but also for harvesting fuel-wood. Both are simple in design and can be manufactured locally from available low-cost materials of construction.

Contact: Mr Erling Fosser, Englekorveien 91, N-1750 Halden, NORWAY, telephone office +47-9-180411, telephone home +47-9-186397, telefax +47-9-187437.

WASTELAND FORESTRY DEVELOPMENT

Mr M S R Prem Kumar
Mr Tota M. Naidu
YCO, INDIA

A system for the management of wasteland, and the development of forestry, based on local participation and charity.

A management system for reforestation projects, based on local participation. It involves tree nurseries, transfer of plants and seedlings to farmers, and the engagement of jobless young people.

Trees are integrated into indigenous agricultural systems, where villagers are responsible for the care and management. The role of the foresters have changed into offering technical support and advice. Villagers are informed of the options and are encouraged to participate in deciding what is appropriate and affordable. The basic principle is 'learning by doing'

The Youth Charitable Organization (YCO) is working as a catalyst. YCO has organized farmers to transport tree seedlings from Government nurseries and plant them on their own land. From this YCO has learned that:

- Tree nurseries are needed close to the villages.
- Seedlings must be ready in time for the monsoon, which is the peak planting period.
- Government nurseries must supply the species of trees that the farmers need and want.

Contact: Mr M S R Prem Kumar, Youth Charitable Organization (YCO), P.O.Box 3, 20/14 Urban Bank Street, Yellamanchili 531 055, Visakhapatnam District, Andhra Pradesh, INDIA, telephone +91-891-77 or 122 - request Yellamanchili, telex 0495-274 CHBR IN - attn: PREMKUMAR OF YCO.

ENERGY

The developing countries have their own energy crisis, which is quite different from the much published situation in the industrially developed countries. Rising prices on imported fossil fuel hit much harder there, but the most serious crisis is the lack of fuelwood for cooking and other household purposes. The local supply of this vital commodity is influenced by such major factors as forest and water management, soil erosion and desertification, to mention just a few. A critical shortage has developed, especially in urban areas.

The situation is made worse by the fact that the infrastructure, in terms of education, research and development, is comparably weak in developing countries. Thus there is a serious dependence on imported technology and know-how. The rate of technical, social and economic innovation is slow, renewal is badly needed and there is a real danger that the economy never will catch up.

This is the difficult setting in which the three award winners have been working, but in spite of economic and other limitations their project has qualified as a complete and successful innovation. It is also uniquely adapted to the particular situation and needs in East Africa. What may look very simple is in fact the result of sophisticated thinking and optimization, fine-tuned with respect to the social, economic and technological environment.

The history of the KCJ project is that of a complex interaction between individual entrepreneurs, governments, funding agencies and NGOs. This innovation is a beautiful example of a comprehensive process, covering the whole chain of events from identification of a need, technical development, establishment of a high volume but small scale production, marketing and dissemination.

The history of fuelwood and stove programmes in developing countries has shown how difficult it is to achieve real results, that have a lasting impact, and not merely satisfy the conscience of a donor. It is therefore a great pleasure to be able to endow one of the few successful projects in the energy area, not at least since this highly innovative and entrepreneurial effort was initiated locally and based on an understanding of local African resources.

Lars Kristoferson

CERAMIC COOKING STOVE

PRIZE WINNERS

Mr A. Awory
Mr S. Karekezi
Mr M. Kinyanjui
KENYA

A simple, energy-efficient, locally manufactured portable cooking stove for a single pot.

This is a simple, portable cooking stove for a single pot. It is locally called the "KCJ", which is short for "Kenya Ceramic Jiko". The KCJ is made up of three distinct components:

- The ceramic firebox.
- The thermal insulation and binder collar.
- The metal cladding.

The KCJ has a very high heat transfer efficiency and reduces fuel consumption by about 50% in comparison with traditional stoves, largely due to good heat insulation.

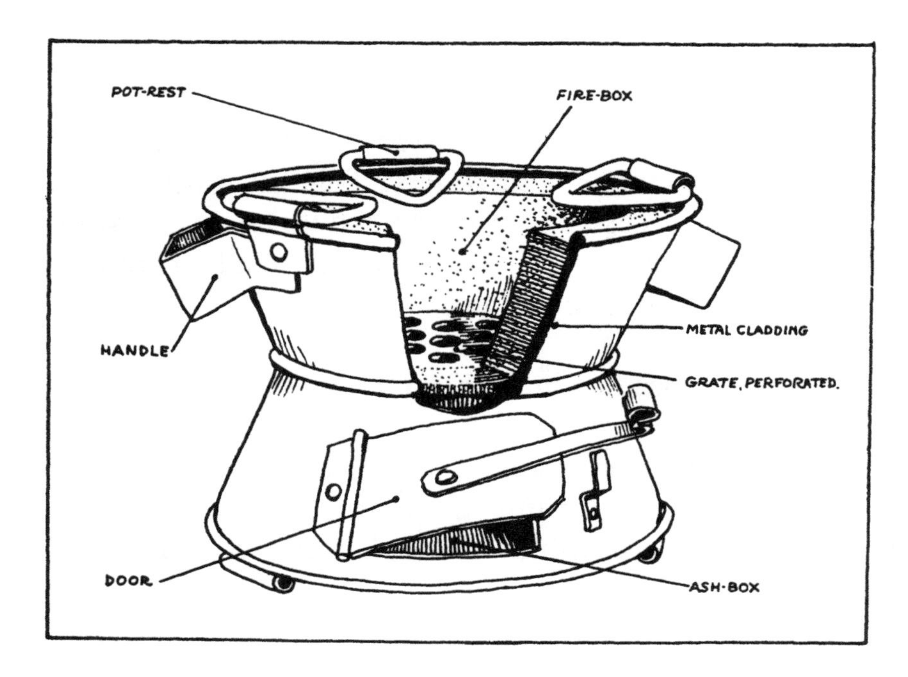

The whole stove is manufactured locally. The ceramic parts are made by village potters, and the cladding is made of sheet metal, recovered from old oildrums etc. The insulating material is mainly local vermiculite. There are two versions of the stove - one for charcoal and one for firewood.

Contact: Mr Stephen Karekezi, Executive Secretary, Foundation for Woodstove Dissemination (FWD), P.O.Box 30979, Nairobi, KENYA, telephone +254-2-566032, telefax +254-2-566032 or +254-2-740524, telex 22912 PUBLIC NRB.

FUEL-SAVING SYSTEM

Mr C.J. Davey
Nairobi, KENYA

A complete system for saving fire-wood in institutional cooking.

The Bellerive Foundation has supported the development of several types of institutional cooking stoves. They are used in hospitals, schools and factories as well as in the villages. Combined with programmes for training the cooks, the new designs have reduced the wood consumption by about 50%. The stoves have the following characteristics:

- The smoke is evacuated by a chimney.
- Stainless steel pots provide fast cooking without burning.
- The rather high stoves makes it easy for the cook to stir the big pot during cooking.
- The food stays warm long after the fire has been extinguished.
- The kitchen remains cool.

The programme includes training in the cutting and proper storage of firewood.

The Foundation has also supported the development of Green Islands, i.e. protected areas, which have been planted with tree seedlings in order to produce firewood for the village.

Contact: Mr Christopher J. Davey, Regional Director, Bellerive Foundation, P.O.Box 42994, Nairobi, KENYA, telephone +254-2-720274 or +254-2-726740, telefax +254-2-726547.

ROTATING PRISM SOLAR WALL

Professor D. Faiman
Ben-Gurion University
ISRAEL

Rotable vertical prismatic columns, which absorb solar heat during the day and radiate it into the house during the night.

A passive solar space-heating system, consisting of a row of rotable vertical columns, each with a triangular equilateral cross-section and flat sides.

The three sides of the columns all have different functions. One side is black, and faces the outside during the day, when it absorbs solar heat. During the night the black surface is facing the inside of the house and radiates the stored heat into the room. The second side is covered with insulating material, and faces the outside during cold winter nights. The third surface is painted according to the aesthetic preferences of the inhabitants. A recently developed programme is used for computing the optimal prism dimensions for a given climate.

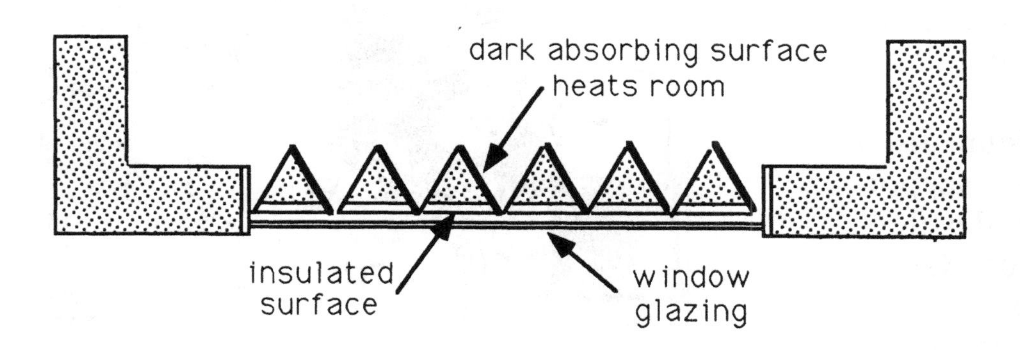

ROTATING PRISM WALL [PLAN]
(in its nighttime heating orientation)

Contact: Professor David Faiman, Center for Energy and Environmental Physics, Jacob Blaustein Institute for Desert Research, Ben-Gurion University of the Negev, Sede Boqer Campus, 84993 ISRAEL, telephone +972-57-565063, telefax +972-57-555058.

SOLAR CANDLE MAKING

Dr P.B.L Chaurasia
CAZRI
Jodhpur, INDIA

A small machine for melting paraffin wax for candle making, based on a flat solar collector as source of energy.

A solar energy driven machine for making candles without the use of electricity. The candles are made of paraffin wax and a few cheap and harmless chemicals. The melting of the raw material is done without the use of conventional fuel, and the machine is very easy to use.

Contact: Dr Pramod Behari Lal Chaurasia, Scientist - SG, Division of Energy Management and Product Processing, Central Arid Zone Research Institute (CAZRI), Jodhpur - 342 003, Rajasthan, INDIA, telephone +91-291-23986, telex 552-218 CZRI IN.

LOW ENERGY HOUSE

Dr John Twidell
ESU
Glasgow, UK

A well insulated house for high latitudes with wind generated electricity and a passive solar conservatory for heating.

This is a low-energy input house for high latitudes. It is heated entirely by renewable energy. Heat is conserved by good insulation of the building and by forced air ventilation with a heat exchanger between the exhaust and incoming air.

Electricity is supplied from a 10kW Aerowatt wind turbine generator to a 2000 litres water tank for heat storage and distribution throughout the house. An L-shaped conservatory on the south and west sides of the house acts as a passive solar collector for heated air.

Reference and owner of the house: Dr Kevin Woodbridge, The Bird Observatory, North Ronaldsay, Orkney, UK.

Contact: Dr John W. Twidell, Energy Studies Unit (ESU), University of Strathclyde, Glasgow G1 1XQ, Scotland, UK, telephone +44-41-5524400 ext. 3307, telex 77472 UNSLIB, telefax +44-41-5521416.

SAVING COOKING ENERGY

Dr Felix A. Ryan
Ryan Foundation
Madras, INDIA

Several inventions relating to fuel for cooking in a small village, making use of twigs, foliage and food waste.

A number of inventions referring to the use of agricultural and silvicultural solid wastes as fuel for cooking in individual households. The inventions explain economic use of these limited resources. They include:

- A PUMP COOKER.
- A system for making FUEL BRIQUETTES out of twigs and foliage.
- A BIOGAS PRODUCER with 3 drums.
- An open air COOKING PLATFORM for community cooking.

Contact: Dr Felix Augustine Ryan, President, Ryan Foundation, 8 West Mada Street, Srinagar Colony, Saidapet, Madras 600 015, INDIA, telephone +91-44-411993.

WATER-DRIVEN POWER GENERATOR

Mr S. Sudarshan
Basaveswaranagar
Bangalore, INDIA

This power generator can be seen as a bucket elevator used for generating energy from a difference in water levels.

This is an alternative to a conventional water turbine. It consists of a simple bucket elevator which is operated backwards, i.e. water is filled into the buckets at the upper water level, and the buckets are emptied at the lower level. The upper or lower elevator shaft is used for driving e.g. an electrical power generator.

Contact: Mr S. Sudarshan, Project Consultant, 7 Vivekananda Road, I-G Cross, 8th Main, SBI Staff Colony, 3rd Stage, 4th Block, Basaveswaranagar, Bangalore 560 079, INDIA, telephone +91-812-356681.

BIOGAS PRODUCER

Dr R. Rabezandrina
Antananarivo
MADAGASCAR

A biogas producer designed for the cooking needs of a family of up to eight adults.

A bio-digester of Chinese type with a production capacity large enough to provide a family of 8 adults with biogas for cooking and other household needs. Each unit has a total volume of 7.5 cubic metres, of which 2.5 is the gas storage dome.

The bio-digester is fed with approximately 40 kilograms per day of fresh animal waste material.

Contact: Dr René Rabezandrina, Département Agriculture de l'Université de Madagascar, B.P. 175, Antananarivo, MADAGASCAR, telephone +261-2-31732.

EVENING LIGHT FROM SPVs

Dr Sanjit Roy
SWRC
Tilonia, INDIA

Units, consisting of solar photovoltaic panels and batteries, to provide light for night classes etc. in rural areas.

Indian villages, out of reach of centralized electricity supply, have been provided with solar photovoltaic (SPV) units. A typical unit consists of one 30 Watts SPV panel, one deep cycle battery, three tube lights and accessories. Such units soon become much more economic than e.g. kerosene lamps.

The first 30 units have been used for providing reading light in evening and night classes for children, who had to look after the family's sheep all day. The school teachers were trained by the Social Work and Research Centre (SWRC) to carry out the maintenance of the unit.

Contact: Dr Sanjit (Bunker) Roy, Director, Social Work and Research Centre (SWRC), Tilonia 305 816, Madanganj, Ajmer District, Rajasthan, INDIA, telephone +91-141-3016.

THE ROCKET COOKING STOVE

Dr L.D. Winiarski
Corvallis
Oregon, USA

A virtually smoke-free stove, which is distributed in the form of a do-it-yourself plan.

The Rocket Stove is available in the form of a do-it-yourself plan. It can be made as separate components and assembled on site. The components are moulded or cast from insulating, heat-resistant material such as adobe, perlite, vermiculite or pumice, and reinforced with chicken wire or fibres. The inner channels, which are exposed to flames, have a sandy clay or pumice liner.

Contact: Dr Lawrence Decker Winiarski, 24395 Starr Creek Rd., Corvallis, Oregon 97333, USA, telephone +1-503-753-4921.

SELF-CONTAINED and MOBILE HEALTH-CARE MODULES

Dr P.G. Virapin
ATDA
Montpellier, FRANCE

Mobile health-care modules, which are powered from solar collector panels, and and thus self-contained to a high degree.

This is a concept called Module Autonome Polyvalent or MAP for short. It is a series of multi-purpose mobile and self-contained modules to be used in human health care. MAP is mounted on a lorry and powered with renewable energy, using solar collector panels.

The essence of this innovation is that the modules are powered by solar energy and thus independent of other sources of energy. They are mounted, temporarily or permanently, on rugged lorries, and are thus extremely mobile. Several modules have been developed, such as a 360 litres cooled storage for vaccines, a 45 litres sterilizer and an atomizer for anti-parasitic agents.

Contact: Dr P.G. Virapin, President, Action Technologique pour le Développement Associé (ATDA), 15 avenue de Maurin, F-34000 Montpellier, FRANCE, telephone +33-67-633952 or +33-67-587877, telefax +33-67-610106 or +33-67-222899.

BLOCK-MAKING PRESS

A team of three
Developm. Altern.
New Delhi, INDIA

A manual press for making soil building blocks, including an information package with practical guidance and advice.

This innovation is the result of team-work carried out by Mr S. Valmeekinathan, Mr Aromar Revi and Mr Sanjay Prakash.

A manually operated block-making press that compacts two soil blocks at a time. The dimensions of the blocks are 230-115-80 millimetres. Five workers can produce 1200 such blocks a day. The press has been named BALRAM. The concept includes not only the press but also an information package containing data about soil testing, soil selection, mix stabilization, block economics, wall and roof design, etc.

Contact: Dr Ashok Khosla, President, Development Alternatives, B-32 Institutional Area, New Mehrauli Road, New Delhi 110 016, INDIA, telephone +91-11-665370, telex 031-73216 DALT IN.

SOLAR HEAT STORAGE

Dr N.M. Nahar
CAZRI
Jodhpur, INDIA

A system for producing warm water with solar energy collectors, including subsequent overnight storage.

A system for solar heating of water and storage of the hot water overnight. All components can be made locally by village artisans.

The device consists of a rectangular galvanized steel tank, which performs the dual function of absorbing heat and storing heated water. The tank is encased in a mild steel tray. The top surface is painted black and is provided with two glass covers.

The same inventor has also designed a small and portable 'Collector cum Storage' device of a more sophisticated design.

Contact: Dr N.M. Nahar, Scientist - SG, Department of Energy Management and Product Processing, Central Arid Zone Research Institute (CAZRI), Jodhpur 342 003, Rajasthan, INDIA, telephone +91-291-23986, telex 552-218 CZRI IN.

NON-TRACKING SOLAR COOKER

Dr N.M. Nahar
Jodhpur, INDIA

A device for cooking with solar heat, suitable for families or for institutions.

A small portable cooker with tiltable solar energy absorbers. The efficiency is such that it can be used also during the Indian winter.

The cooker consists of a double-walled box with insulating material between the walls. The inner wall is painted black in order to absorb solar energy. The box is covered with a sheet of glass in order to keep the heat inside, where the cooking utensils are placed. The box is provided with two reflecting lids.

Contact: Dr N.M. Nahar, Scientist - SG, Department of Energy Management and Product Processing, Central Arid Zone Research Institute (CAZRI), Jodhpur 342 003, Rajasthan, INDIA, telephone +91-291-23986, telex 552-218 CZRI IN.

SOLAR DRIER

Dr P.C. Pande
CAZRI, Jodhpur
INDIA

A multi-purpose solar dryer for sensitive food products, with automatic temperature regulation.

A solar drier for sensitive food products such as fruit and vegetables. The drier features an automatic air temperature regulation, which prevents over-heating. A regulating baffle in the chimney controls the air circulation. The food products are dried on removable wire mesh trays or, as in a second version, in tiltable boxes, provided with glass lids. The device can be used also for water heating.

A modified multi-purpose device can be used both for drying and cooking.

Contact: Dr Piyush Chandra Pande, Head of Division, Division of Energy Management, Engineering and Product Processing, Central Arid Zone Research Institute (CAZRI), Jodhpur 342 003, Rajasthan, INDIA, telephone +91-291-23986, telex 552-218 CZRI IN.

WATER

Most nominations in this target area relate to different aspects of the provision of drinking water, such as pumping, purification, sterilization etc., but also to solar distillation systems. Another group of nominations refer to sewage and waste water treatment. Two nomination are systems for rainwater catchment and storage, including the management of water resources.

Several ways of powering simple water pumps have been developed. One nomination is a treadle pump, and one water-elevating system is powered from a revolving wicket gate.

Many nominations are solar driven devices. One example of that is a multiple-wick solar still that produces drinking water from inferior forms of raw water. The ultraviolet radiation system for sterilization of water is powered from a photo-voltaic panel.

The two sewage treatment systems are based on biological treatment. One of them employs complete ecological systems.

The prize winner has developed a system for the maintenance of village hand-pumps. It is mainly a social innovation, even if it has a strong technical component. It implies the appointment of a suitable villager to become the local handpump mechanic (HPM). This person is given a thorough practical training and is provided with proper tools and transport facilities. Thus competence, resources and responsibility are removed from local government, decentralized and distributed among the villages. Placing the handpump mechanic in the village itself makes him/her accountable to the local community. It provides young people with incentives and makes them feel useful and rewarded. The system furthermore encourages preventive maintenance as opposed to repair after break-down. It is worth mentioning that lack of reliability, traditionally has been the most important single factor hampering the full utilization of village handpumps. Maintenance has been the stumbling block in attempts to improve the water-supply and the sanitation in the villages.

The 1990 IDEA Award winner has been selected for his whole-hearted engagement in the organization and practical implementation of a reliable and user-oriented water-supply system for Indian villages.

Bertil Hawerman

HAND-PUMP MAINTENANCE

PRIZE WINNER

Mr Sanjit Roy
SWRC
Rajasthan INDIA

An organization with many social incentives for decentralized and regular maintenance of village hand pumps.

The Handpump Mistry (HPM) is a simple but effective system for maintenance of handpumps (India Mark II) in rural areas. Mistry is the Hindi word for Mechanic. The Mistry is found in every village in India, repairing tractors, bullock carts, diesel and electric pumps and agricultural implements, without having gone through any formal training. The HPM service includes every part of the pumping system, below ground level as well as above.

This innovation is of a social character rather than technical. It implies the nomination of a suitable villager to become the local HPM. The Mistry receives a thorough practical training, is provided with proper tools, and is allotted 30 to 40 pumps within a radius of 5km from home. The Mistry gets a monthly payment, provided all his pumps are in working order. Thus competence and resources are decentralized and distributed, thereby considerably reducing government expenditure.

HPM has the potential to achieve the following:

- Generate self-employment in the villages ona large scale, especially among young people without work.
- Reduce the costs for pump repair and maintenance.
- Reduce the dependency on government and international agencies.
- Demystify technology.
- Promote personal participation.
- Give respectability, credibility and recognition to local people's skills and knowledge.
- Remove the linking between school degrees and practical jobs in the village.
- Develop self-respect in poor communities, which is more important than self-reliance.

Contact: Mr Sanjit (Bunker) Roy, Director, Social Work and Research Centre (SWRC), Tilonia 305 816, Madanganj, Ajmer District, Rajasthan, INDIA, telephone +91-141-3016.

VILLAGE WATER GATE

Mr Felix A. Ryan
Madras, INDIA

A water-elevator driven by a gate in the form of a revolving door.

This very simple water-lifting device consists of two main parts, i.e. a water-elevator and a revolving wicket gate.

The water-elevator consists of two rope chains, with PVC or canvas buckets, with a shaft at the top and another at the bottom of the well.

Energy to drive the water-elevator is produced by the gate in the form of a revolving door, through which the villagers have to pass when they go to the well to fetch water, as well as when they leave.

School children and members of the village community are asked to use the gate as a thoroughfare. For every full turn of the gate a bucket of water empties itself into a water-storage tank. From the tank, water is conducted to the individual households through bamboo or hose pipes.

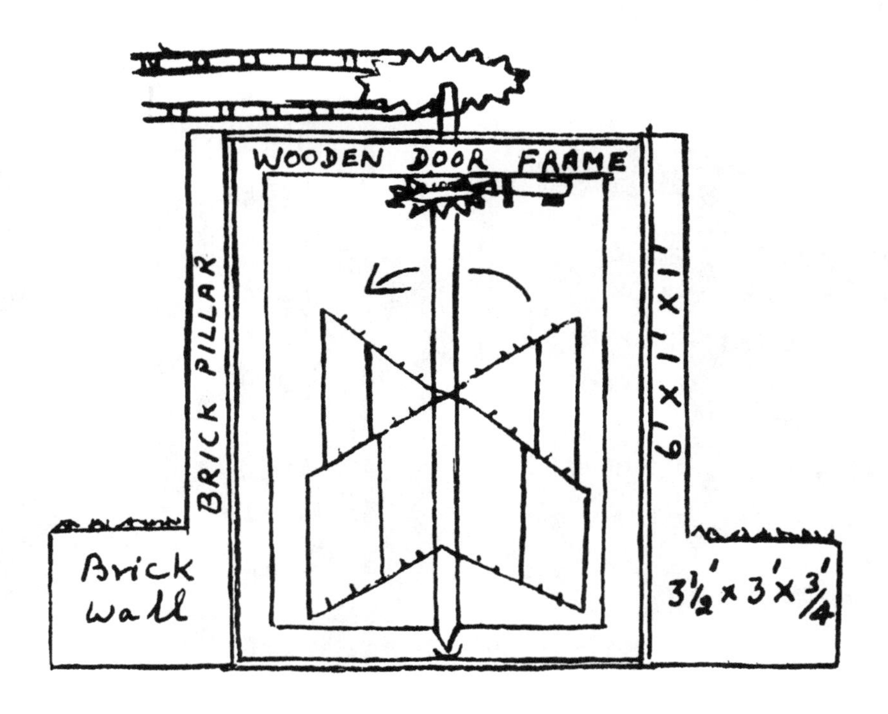

Contact: Mr Felix Augustine Ryan, President, Ryan Foundation, 8 West Mada Street, Srinagar Colony, Saidapet, Madras 600 015, INDIA, telephone +91-44-411993.

‘TANKA’ RAINWATER HARVESTING

Mr N.S. Vangani
Dr K.D. Sharma
CAZRI, Jodhpur
INDIA

‘Tanka’ is a system for rainwater harvesting in a desert climate, consisting of an underground tank with appropriate catchment.

‘Tanka’ is the name given to a system for rainwater harvesting in the Indian desert, but was originally a local term for an underground cistern. It has a circular cross-section, because this shape is stronger and more economical to build. The catchment area has to be planned properly and has to meet the following criteria:

- Produce adequate run-off in order to meet the need of drinkingwater over the whole year.
- Cause a minimum of soil damage and loss.
- Cause a minimum of evaporation and pollution.
- Proper management and maintenance.

The innovators strongly recommend individual ownership. In the past, communal ownership has failed for several reasons.

Contact: Mr N.S. Vangani or Dr K.D. Sharma, Central Arid Zone Research Institute (CAZRI), Jodhpur 342 003, Rajasthan, INDIA, telephone +91-291-23986, telex 552-218 CZRI IN.

TREADLE PUMP

Mr Carl Bielenberg
Washington, DC
USA

A simple treadle pump for small farm irrigation. It can easily be manufactured locally, mainly in sheet steel.

A simple foot-operated pump for irrigation, that can be manufactured locally from easily available material, such as sheet steel, using low cost sets of jigs and fixtures.

The original concept was developed in Bangladesh by Gunnar Barnes with support from Rnagpur-Dinajpur Rehabilitation Service and Lutheran World Federation. While the original pump was for suction lift only, the more recent design is for pressure and suction lifting, in order to facilitate pumping from surface water sources and for piped irrigation in sandy soils. In Mali the pump is manufactured by artisans, with technical and marketing assistance provided by Association d'Etudes de Technologies Appliquées et d'Amenagement en Afrique (AETA). Manufacturing is beginning also in Senegal and Cameroon.

This treadle pump is a twin cylinder device that uses the legs of one or two adults in comfortable walking motion. It can deliver approximately seven cubic metres of water per hour depending upon the total head required.

This treadle pump is very versatile, but has so far been used mainly for the irrigation of small market gardens and vegetable farms.

The Treadle Pump

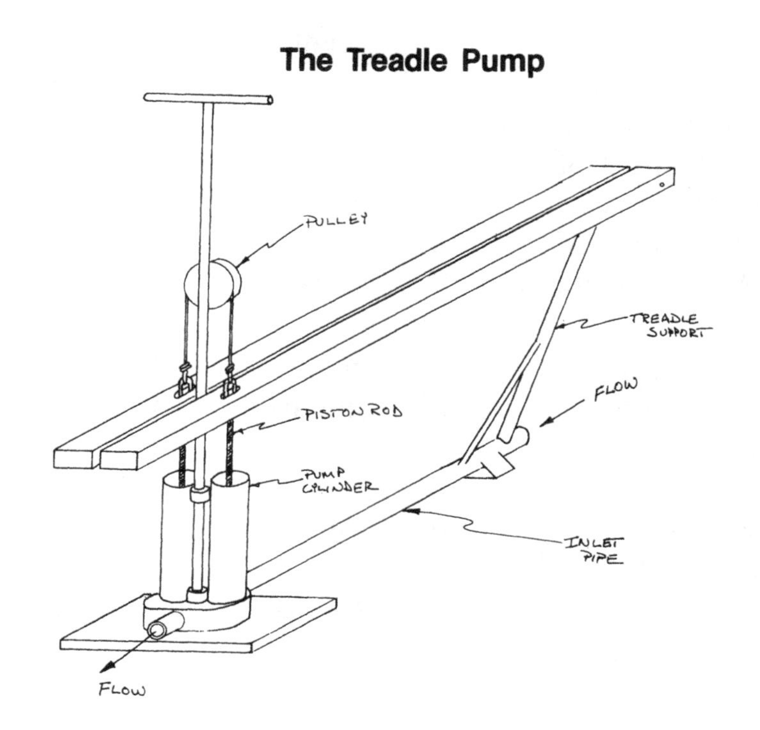

Contact: Mr Carl Bielenberg, c/o Appropriate Technology International (ATI), 1331 H Street NW, Washington DC 20005, USA,.telephone +1-202-879-2900, telex 64661 ATI, telefax +1-202-628-4622.

WATER-RESOURCE MANAGEMENT

Dr P. Khanna
Nagpur, INDIA

A holological approach to water-resource management in semi-desert areas.

A holological approach to sustainable water-resource management, with special emphasis on analysis of the supportive capacity of the environment and proper implementation.

Water is a regenerative natural resource. The supportive capacity of a regional environment determines the quantity of water available, and its assimilative capacity determines the quality. Water-resource management, based on carrying capacity, thus requires consideration of these factors and their interaction.

The following activities are particularly important for a holological analysis of water-resource management:

- Hydro-geology and source identification.
- Water and wastewater (for reuse) quality assessment and treatability studies.
- Engineering analysis and design.
- Preparation of software for operation and maintenance.
- Human resource development.
- Preparation of water-balance statements and detailed project reports.
- Environmental impact assessment.

The analysis comprises not only available water-resources, but also the capacity of the environment to absorb residuals and wastes from all the human activities in the area.

The implementation of the programme has included effective participation from the beneficiaries, who have been actively involved in planning, construction, sustained operation and maintenance of the individual projects. This has created awareness, motivation and community spirit.

Contact: Professor Dr P. Khanna, Director, National Environmental Engineering Research Institute (NEERI), Nehru Marg, Nagpur 440 020, INDIA, telephone +91-712-523893, telefax +91-712-523893, telex 953-0715-233 NERI IN.

MULTIPLE-WICK SOLAR STILL

Dr G.N. Tiwari
CES / IIT
New Delhi, INDIA

A multiple-wick solar still, constructed in fibre reinforced plastics, with wicks in black jute.

The multiple-wick solar-distillation unit consists of a frame made of fibre-reinforced plastic (FRP). It encloses a sheet of glass, covering several interspaced layers of jute and sheets of polyethene. Both the jute wicks and the sheets are black in order to absorb solar radiation. Each pair of wick and sheet is shorter than the next lower pair, thus exposing the ends to the sunshine. The upper ends of the jute wicks dip into a reservoir of raw water. Desalinated and purified water is condensed on the sloping glass cover and collected at the lower end of the frame.

The inventor claims the following advantages with his design:

- The condensing glass cover and the evaporative black surfaces are parallel.
- The whole frame can be placed at any slope in relation to the horizontal plane, in order to optimize its angle towards the prevailing sun.

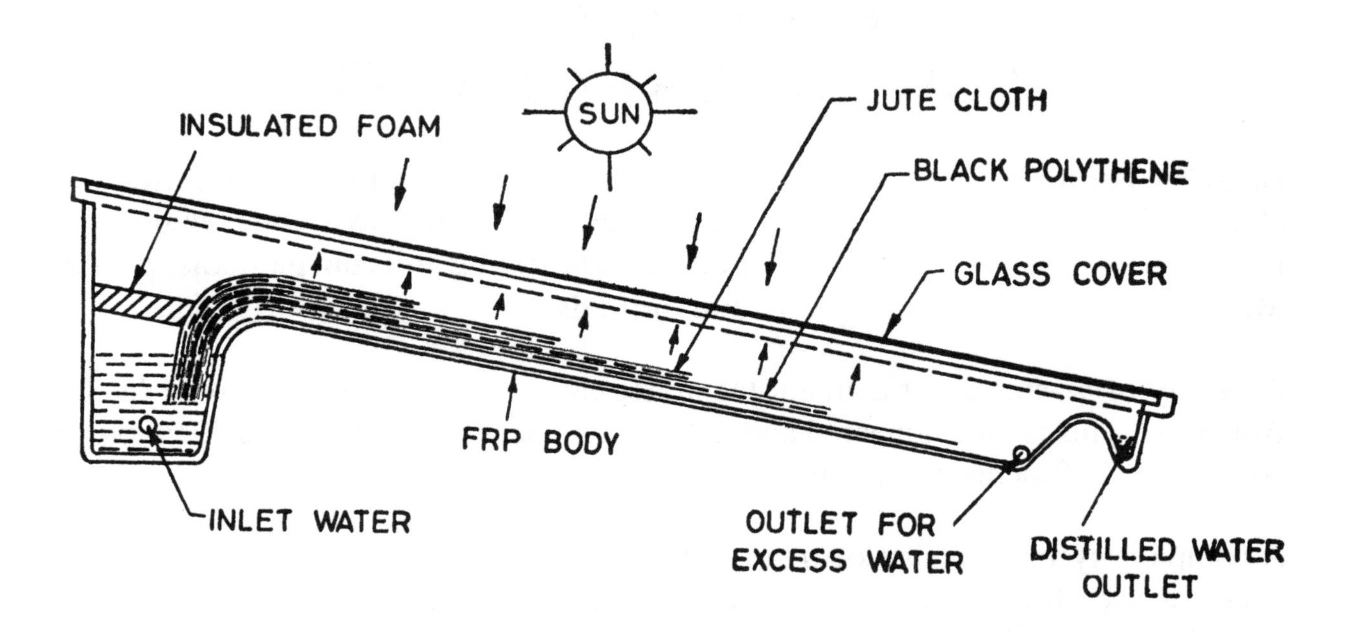

Contact: Dr Gopal Nath Tiwari, Assistant Professor, Centre of Energy Studies (CES), Indian Institute of Technology (IIT), Hauz Khas, New Delhi 110 016, INDIA, telephone +91-11-654054, telefax +91-11-954321, telex 31-73087 IIT IN.

UV DISINFECTION OF WATER

Dr Otto F. Joklik
Vienna
AUSTRIA

A water purification and disinfection plant, using ultraviolet radiation for producing safe drinking water.

A portable water cleaning and disinfection plant, that produces potable water, and which consists of the following main components:

- An 80 mesh stainles steel filter for removal of suspended solid impurities.
- An activated carbon filter for removal of chemical impurities.
- An ultraviolet radiation treatment, powered from a battery or photovoltaic solar panels.

The plant is operated entirely with solar energy.

Contact: Dr Otto F. Joklik, Gersthoferstrasse 120, A-1180 Vienna, AUSTRIA, telephone +43-222-473122.

ECOLOGICAL SEWAGE TREATMENT

Dr John Todd
Ocean Arks
Falmouth
MA, USA

A system for treatment of concentrated waste water in an artificial ecological system comprising microorganisms, plants and animals.

A family of wastewater treatment technologies, which have been named 'Solar Aquatics'. This is an engineered and tailor-made ecosystem for the purification of concentrated effluents, such as municipal sewage.

A complete system consists of treatment tanks and green plants under cover. The conditions are fully controlled and replicated. The input consists of wastewater, sunlight and aeration energy.

Contact: Dr John Todd, The Center for the Protection and Restoration of Waters,.Ocean Arks International, 1 Locust Street, Falmouth, MA 02540, USA, telephone +1-508-540-6801, telefax +1-508-548-4359.

DRINKING-WATER PURIFICATION

Dr Fouad Abosamra
SSRI
Damascus, SYRIA

A mobile stand-alone plant for the purification of polluted water into drinking water.

A mobile system for the purification of polluted water into drinking water, comprising the following steps:

- Lamellar settling tank.
- Pressure sand filter.
- Treatment with activated carbon.
- Treatment with chlorine.

The plant is powered by a diesel engine and an electricity generator, which drives the main pump.and auxiliaries.

Contact: Dr Fouad Abousamra, Scientific and Research Center (SSRC), Damascus, SYRIA.

BIOLOGICAL SEWAGE TREATMENT

Mr Elias Khoudary
EIC
Aleppo, SYRIA

A small and compact activated sludge plant for municipal sewage. It is manufactured and delivered as one complete unit.

A small package plant for the biological treatment of sewage, including intensive aeration. The air is injected through open pipes as well as through ceramic diffusers, resulting in small air bubbles. The air injection causes the tank content to flow and mix in a controlled way.

The plant is pre-fabricated and delivered as a complete unit. It is designed for a population of 200 to 2000 persons.

Contact: Mr Elias Khoudary, Engineering & Industrial Contracting (EIC), P.O.Box 6611, Riayat al-Shabab, Aleppo, SYRIA, telephone +963-21-462574 & 448976.

INDEXES

It is our intention and hope that this book will be used as a source of information for people working in, or for, developing countries. In order to make it more practical as a reference book to appropriate technology, we have provided it with several indexes. Besides the subject index, there is also an index of inventors and authors, of institutions and of national origin.

SUBJECTS

Some of the original Target Areas have been split up into smaller areas in order to facilitate the use of this index. A number of items have been placed in more than one area, and have been provided with multiple entries.

Agriculture

Agro-Forestry

Aquaculture

Energy

Fishing

Food

Forestry

Health Care

Housing

Water

INVENTORS and AUTHORS

INSTITUTIONS

NATIONS

www.ingramcontent.com/pod-product-compliance
Lightning Source LLC
Jackson TN
JSHW060309140125
77033JS00021B/630

* 9 7 8 1 8 5 3 3 9 1 0 2 6 *